SECOND EDITION

PRELIMINARY EDITION
SOFTWARE MANUAL

POWER SYSTEM ANALYSIS AND DESIGN

J. Duncan Glover
Failure Analysis Associates, Inc.--Electrical Division

Gareth Digby
University College of Swansea
Swansea, Wales
UK

PWS Publishing Company
Boston

PWS PUBLISHING COMPANY
20 Park Plaza, Boston, MA 02116-4324

PWS Publishing Company is a division of Wadsworth, Inc.

I(T)P ™
International Thomson Publishing
The trademark ITP is used under license.

Printed and bound in the United States of America by Financial Publishing Company.

2 3 4 5 6 7 8 9 10 - - 98 97 96

ISBN 0-534-93962-7

CONTENTS

PREFACE

Digital computer programs are used extensively in the analysis and design of electric power systems. With personal and desktop computers, users have the convenience of access at any time or place. This software package contains a set of programs that enable engineering students and engineers to conduct power system analysis and design studies. The data-handling and number-crunching capabilities of the PC allow students and professionals to work on more difficult and realistic problems and make the PC an innovative tool in the learning process.

The floppy disk included with this software package can be run on the IBM PC or any compatible personal computer that is able to run Microsoft Windows 3.1 or higher. The set of programs stored on the disks can be used to solve the personal computer problems given in the accompanying text *Power System Analysis and Design With Personal Computer Applications,* Second Edition, published by PWS Publishers in 1994. Users can also run the programs for their own power system studies.

This software package is primarily intended for education, not for solving large dimensional power systems. Thus the power flow, short circuit, and transient stability programs are capable of handling up to 100-bus power systems.

Each of the following gentlemen wrote one of the programs in this software package, as graduate students under the supervision of J. Duncan Glover.

Jalil Elmernissi - LINE CONSTANTS (CHAPTER 5)

Jean Y. Ayoub   - POWER FLOW (CHAPTER 7)

Terry J. Fundak - SYMMETRICAL SHORT CIRCUITS (CHAPTER 8)

Mr. Fundak also assisted in the writing of SHORT CIRCUITS (CHAPTER 9).

Dr. Gareth Digby has been converting the programs to a Microsoft Windows-based application. Currently CHAPTER 2 (Matrix Operations) and CHAPTER 7 (Power Flow) are Windows-based.

It is hoped that this software package will be useful to students and professionals in their daily studies of power system analysis and design. For those who are studying from the text *Power System Analysis and Design With Personal Computer Applications,* the software can solve practical, real-world problems to supplement the theory, and can open the door to design orientation in power systems.

J. Duncan Glover

Gareth Digby

1. INTRODUCTION

This manual gives user instructions for the computer programs stored on the floppy disk that accompanies this software package. No programming knowledge is required on the part of the user to run these programs.

The programs are written to run on an IBM or compatible PC. The basic requirements are that the computer should be able to run Microsoft Windows 3.1 and have one 3½ inch, 1.44M floppy disk drive and a hard disk with 1,120,515 bytes of available space. A Microsoft Windows compatible printer is also needed for outputs.

The disk that accompanies this software package contains all the program and data files. The programs are described in further details in Chapters 2 - 13 of this manual. Program flow charts are given in Chapter 14, and sample single-line diagrams for power system analysis and design studies are given in Chapter 15. Five student design projects are given in Chapter 16. The background and theory for the programs are given in the text *Power System Analysis and Design with Personal Computer Applications*, Second Edition, published by PWS Publishers, Boston, 1994.

There is no software for Chapters 1, 4, 10, and 11 of the text. In order to make software chapters correspond with chapters in the text, Chapters 4, 10, and 11 of this manual are intentionally left blank.

INSTALLATION

The disk that accompanies this software package is write-protected in order that you have a copy of the programs that cannot easily be erased or unintentionally modified. It is first necessary to install the software on your computer's hard disk before you can run programs. Programs cannot be run directly from the floppy disk.

To install the software on your hard disk, ensure that Microsoft Windows 3.1 or above is executing on the computer. Insert the floppy disk into the disk drive. Run the PROGRAM MANAGER.* Select the FILE menu, and chose the RUN... item. At the prompt, type: A:SETUP

(If the floppy disk is in drive B, type: B:SETUP)

The SETUP application will then load and prompt for the destination hard disk drive and directory. The default destination drive and directory are C:\PSAD. If you wish to install the software in another drive and directory, type your selection at the prompt. (The application will automatically detect the directory holding the Microsoft Windows operating system.)

---

*   If you have Norton Desktop, it is not necessary to run the Program Manager.

The SETUP application will install the software and create a window within the PROGRAM MANAGER that holds application icons. The window is entitled POWER SYSTEM ANALYSIS AND DESIGN. All the applications and programs are launched from this window. Either single-click on an icon and then select the FILE menu and OPEN item, or double-click on an icon.

The applications CHAPTER 2 (Matrix Operations) and CHAPTER 7 (Power Flow) are Microsoft Windows based applications. Their operation is described in greater detail in Chapters 2 and 7 of this manual.

The programs CHAPTER 3 (Symmetrical Components), CHAPTER 5 (Line Constants), CHAPTER 6 (Transmission Lines: Steady State Operation), CHAPTER 8 (Symmetrical Short Circuits), CHAPTER 9 (Short Circuits), CHAPTER 12 (Transmission Line Transients), and CHAPTER 13 (Transient Stability) are Microsoft DOS based programs. (These programs are still launched in a similar manner to Microsoft Windows based applications, by double clicking on the icon, etc.)

DATA FILES FOR MICROSOFT DOS-BASED PROGRAMS

Data files are used with CHAPTER 5, CHAPTER 8, CHAPTER 9, CHAPTER 12 and CHAPTER 13 programs. All these data files have the extension ".DAT".

Each program that uses data files has a "display the data file" selection. Therefore, the program user can display data either on the monitor or with the printer while executing the program. It is not necessary to directly examine the contents of a data file.

For each program that uses data files, the program user can store the data for three to five separate cases. For example, the data file that is used to store the input data for Case 1 of CHAPTER 5, (Line Constants), is called LC1.DAT. The input data file for Case 2 (3, 4, 5) has a similar name, LC2.DAT, except that the number 2 (3, 4, 5) is used. Thus you can store the CHAPTER 5 input data for five different transmission lines.

Each time you update the input data files for a program, the program writes over the data files. That is, it replaces the data files in your directory with the updated ones. In this way, the total disk space used for storage does not increase. Also, you don't have to erase any data files.

SAVING CASES FOR THE MICROSOFT DOS BASED PROGRAMS

For each DOS based program that uses data files, you can save three to five cases in the program directory. If you have more cases that you want to save, or if you want a backup copy of the data for a particular case, you can COPY the files to another directory. Or COPY the files under a new name. The Microsoft Windows FILE MANAGER application can be used for this purpose. (Do NOT MOVE the files to a new directory. Also, do NOT RENAME the files in the program directory. The programs will not execute properly if the files are missing.)

HELPFUL HINTS FOR MICROSOFT DOS-BASED APPLICATIONS

ENTERING DATA

When entering data for any program, it is helpful to have a single-line diagram of your power system directly in front of you. As such, you can see the system interconnections on a single-line diagram as you enter data. Also, all power system data should already be converted to per unit on a common system base.

ENTERING ZEROS

To enter a 0 (zero) during data entry, you do not have to type in the number 0. Just press the ← key. It is faster this way.

CORRECTION OF ERRORS BEFORE PRESSING

Most of the programs give you options for changing the input data. However, if you make a mistake in data entry before pressing the key, the input data has not yet been registered in the computer. You can use the ← key to backspace and erase the mistake, then enter the correct data.

CTRL PRINT SCREEN

When it is time to print input/output data, the programs will tell you to use the Ctrl Print Screen option if you want to print the data on a printer. Press the Ctrl key and the Print Screen key simultaneously. Also, make sure your printer is turned on. The printer will then print out whatever appears on the monitor. You can use this option at any time. (Problems occur if DOS-based programs try to print to a Postscript printer.)

2. MATRIX OPERATIONS (CHAPTER 2)

This application performs the following functions:

Matrix Addition	$C=A+B$ of two N x M matrices **A** and **B**. The matrices **A** and **B** can be either real or complex.
Matrix Subtraction	$C=A-B$ of two N x M matrices **A** and **B**. The matrices **A** and **B** can be either real or complex.
Matrix Multiplication	$C=A \times B$ of two matrices **A** and **B**. The matrices have dimension N x M for **A**, and M x P for **B**. The matrices **A** and **B** can be either real or complex.
Matrix Inverse	A^{-1} of the N x N matrix **A** whose determinant is assumed to be non zero. Matrix **A** can be either real or complex.
Matrix Transpose	A^{T} of the N x M matrix **A**. Matrix **A** can be either real or complex.
Complex Conjugate	A^{*} of the N x M complex matrix **A**.

The maximum dimension of any matrix is 30. The elements of complex matrices can be entered in either rectangular or polar form. The data and results are displayed on the screen and can be printed out. Also, the data and results can be displayed in either exponential format (similar to scientific format) or fixed-point format (numbers with decimal points).

APPLICATION OPERATION

The application is started in the usual Windows manner by double clicking on the "Chapter 2 Matrix Operations" icon in the **Program Manager** "Power System Analysis and Design" window.

The application presents the user with a main parent window entitled "Matrix Operations" holding three smaller child windows, entitled "Matrix A: Input: real", ."Matrix B: Input: real", "Matrix C: Results: real". See Figure 2a.

The Title Bar at the top of parent window has the usual Windows Control Box button on the left hand side, and the Minimize and Maximize buttons on the right hand side. The Menu bar at the top of the parent window is described later in this chapter. The bottom of the parent window shows a Status bar. The Status bar shows the active child window and the value of the current cell in that child window.

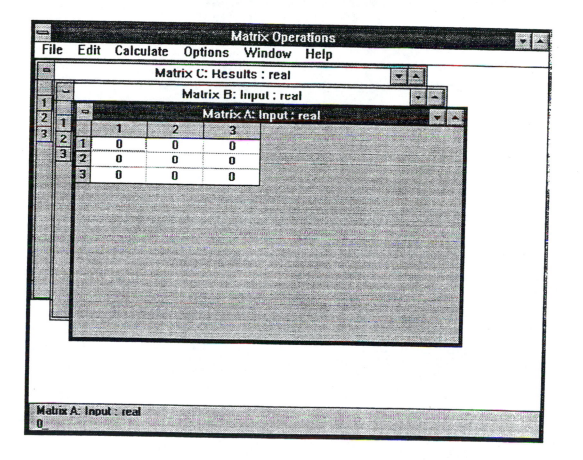

Figure 2a. Parent Window

5

ENTERING DATA

Select either the matrix A or B child window to make it the active matrix child window. Selection can be achieved by either clicking on the child window, or selecting the appropriate item in the Window menu. The cell for the data to be entered into is selected by either clicking on the cell, or using the arrow keys to move around the cells. The contents of the active cell are displayed in the status bar. The contents of the cell can be erased by using either the delete key to remove the entire contents, or the backspace key to remove the left hand character in the string. Values can be entered in the form:

Real	Complex
123.45	123.45+j67.89
1.2345e2	1.2345e2+j6.789e1

MENUS

The Menu bar has the following groups: **File, Edit, Calculate, Options, Window, Help.**

The **File** menu has the following items:

Item	Action	Short Cut Key
New	Clears all three matrices **A**, **B** and **C**	
Open	Opens a file holding values for matrices **A** and **B**, and reads in values.	
Save	Saves contents of matrices **A** and **B** in a file.	
Save As	Saves contents of matrices **A** and **B** in a file. The user is allowed to specify a new file name.	
Print	Prints out contents of matrices **A**, **B** and **C** on default Windows printer.	Ctrl+P
Exit	Exits application.	

The Edit menu has the following items:

Item	Action	Short Cut Key
Undo	Undo last edit operation on active matrix window.	Ctrl+Z
Cut	Cut contents of selected cells in active matrix window to the Clipboard.	Ctrl+X
Copy	Copy contents of selected cells in active matrix window to the Clipboard.	Ctrl+C
Paste	Paste contents of Clipboard into selected cells in active matrix window.	Ctrl+V
Delete	Delete contents of selected cells in active matrix window.	Del
Select All	Select all cells in the active matrix window.	
Matrix Type	Allows type of active matrix to be changed through a dialog box. The matrix type can be either Real or Complex.	
Matrix Size	Allows size of active matrix to be changed through a dialog box.	
Zero Matrix	Fills the active matrix with the value zero in every cell.	
Unit Matrix	Fills the active square matrix with the Unit Matrix.	

The **Calculate** menu has the following items:

Item	Action
Add	C=A+B. Adds matrices **A** and **B**, placing result in matrix **C**.
Subtract	C=A-B. Subtracts matrix **B** from matrix **A**, placing the result in matrix **C**.
Multiply	C=A*B. Multiplies matrix **A** by matrix **B**, placing the result in matrix **C**.
Inverse	Inverts the active matrix (either **A**, **B** or **C**), placing the result in the active matrix.
Transpose	Transposes the active matrix (either **A**, **B** or **C**), placing the result in the active matrix.
Conjugate	Calculates the conjugate of the active matrix (either **A**, **B** or **C**), placing the result in the active matrix.

The **Options** menu has the following items:

Item	Action
Print Format	The format used to print the contents of the cells; can be set to either Fixed or Exponential.

The **Window** menu has the following items:

Item	Action
Cascade	Arranges the three child windows in a cascade within the main window.
Tile Horizontally	Arranges the three child windows tiled horizontally within the main window.
Tile Vertically	Arranges the three child windows tiled vertically within the main window.
Restore All	Restores all child window icons to their original size.
Minimize All	Minimizes all the child windows into icons.
Arrange Icons	Tidies up the arrangement of any child window icons.
Refresh	Refreshes the display.
1	Selects Matrix C: Results as the active child window.
2	Selects Matrix B: Input as the active child window.
3	Selects Matrix A: Input as the active child window.

The **Help** menu has the following item:

Item	Action
About	Lists application authors and shows system statistics.

SAMPLE RUN CHAPTER 2 MULTIPLYING MATRICES A AND B

Select matrix **A** as the active matrix. Set matrix **A** type to complex using the **Edit** menu, **Matrix Type** item. Fill matrix **A** with the unit matrix using the **Edit** menu, **Unit Matrix** item.

Then select matrix **B** as the active matrix. Set matrix **B** type to complex using the **Edit** menu, **Matrix Type** item. Fill matrix **B** with contents listed here:

1.5+j2	1.25+j2.5	4.3+j2
2.4+j1.2	1.6+j1.2	2.4+j3.1
4.3+j1.3	3.7+j2.1	1.7+j2.3

Then multiply matrix **A** by matrix **B**, by using the **Calculate** menu, **Multiply** item. Use the **Window** menu, **Tile Horizontally** item to present the matrices in the form shown in Figure 2b.

CUT, COPY AND PASTE

When copying between matrices within the application, the same number and arrangement of cells must be selected, in the matrix copied from, and the matrix pasted to.

The **Cut, Copy and Paste** actions use the Windows Clipboard as the destination for **Cut** and **Copy**, and the source for **Paste**, operations. Therefore it is possible to copy the contents of any of the matrices into another Windows application that will accept input from the Clipboard. Also values can be pasted from another application into a matrix through the Clipboard. (The values for different columns must be separated by tabs. Rows are separated by carriage returns in the other application. Do not include any spaces.)

The **Edit** menu, **Select All** item selects all the cells in the active matrix. Individual cells can be selected by clicking on them. Groups of cells can be selected by clicking on the first cell, and while holding down the mouse button, dragging across the other cells in the group. An entire row (or column) can be selected by clicking on the row (or column) number. Multiple rows (or columns) can be selected by clicking on the first row (or column) number and while holding down the mouse button, ragging across the other row (or column) numbers required.

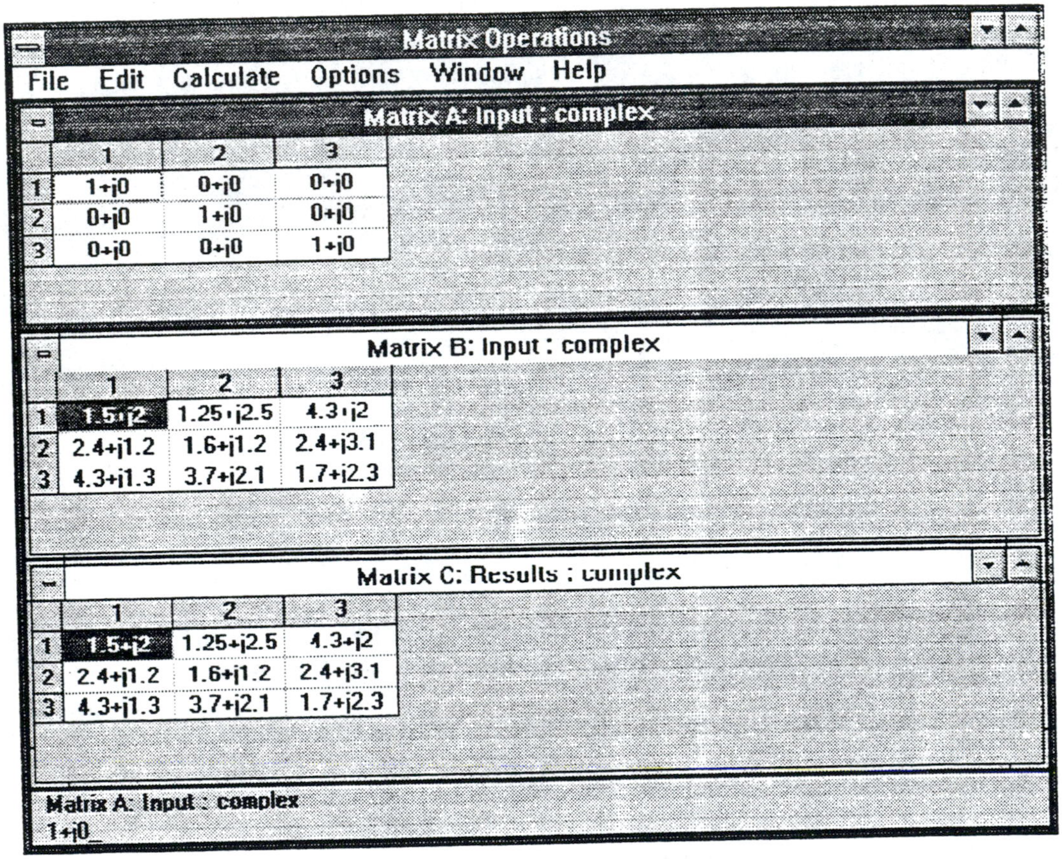

Matrix Operations

File   Edit   Calculate   Options   Window   Help

Matrix A: Input : complex

	1	2	3
1	1+j0	0+j0	0+j0
2	0+j0	1+j0	0+j0
3	0+j0	0+j0	1+j0

Matrix B: Input : complex

	1	2	3
1	1.5+j2	1.25+j2.5	4.3+j2
2	2.4+j1.2	1.6+j1.2	2.4+j3.1
3	4.3+j1.3	3.7+j2.1	1.7+j2.3

Matrix C: Results : complex

	1	2	3
1	1.5+j2	1.25+j2.5	4.3+j2
2	2.4+j1.2	1.6+j1.2	2.4+j3.1
3	4.3+j1.3	3.7+j2.1	1.7+j2.3

Matrix A: Input : complex
1+j0_

Figure 2b. Sample Run Chapter 2

10

3. SYMMETRICAL COMPONENTS (CHAPTER 3)

This program contains the following three symmetrical component subroutines:

(1) SEQVEC - computes the complex sequence vector $V_s = A^{-1} V_p$
for any three-phase complex phase vector V_p.

(2) PHAVEC - computes the complex phase vector $V_p = A V_s$
for any three-phase complex sequence vector V_s.

(3) SEQIMP - computes the 3 x 3 complex sequence impedance matrix $Z_s = A^{-1} Z_p$
A for any 3 x 3 complex phase impedance matrix Z_p.

The A matrix in the above subroutines is the 3 x 3 symmetrical components transformation matrix given by (3.1.8) of the text *Power System Analysis and Design With Personal Computer Applications*. Also, the V vectors in the above subroutines can be three-phase voltage vectors, three-phase current vectors, or any vectors with three complex components.

You can enter the elements of the complex vectors and matrices in either polar or rectangular form. The input/output data can be displayed on the monitor and the printer. Also, you can display the input/output data in either exponential format (similar to scientific notation) or in fixed point format (numbers with decimal points).

The following sample run computes the sequence components of an unbalanced three-phase voltage vector.

SAMPLE RUN CHAPTER 3 SEQUENCE COMPONENTS OF V_p

```
THIS PROGRAM CONTAINS THE FOLLOWING THREE SUBROUTINES:

1: SUBROUTINE SEQVEC(VP,VS)
     SEQVEC COMPUTES THE SEQUENCE VECTOR VS=(AINV)VP
2: SUBROUTINE PHASVEC(VP,VS)
     PHAVEC COMPUTES THE PHASE VECTOR VP=(A)VS
3: SUBROUTINE SEQIMP(ZP,ZS)
     SEQIMP COMPUTES THE SEQUENCE IMPEDANCE MATRIX ZS=(AINV)ZP(A)

 TO USE ONE OF THE ABOVE SUBROUTINES, ENTER AN INTEGER FROM
 1 TO 3 OR ENTER THE INTEGER 4 TO EXIT.

WHICH SUBROUTINE(ENTER AN INTEGER FROM 1 TO 4)? 1


WILL YOU ENTER THE DATA IN RECTANGULAR COORDINATES (Y OR N)? n

DO YOU WANT THE OUTPUT IN RECTANGULAR COORDINATES(Y OR N)? n

DO YOU WANT THE OUTPUT DATA IN EXPONENTIAL FORMAT (Y OR N)? n
```

```
*****INPUT/OUTPUT DATA FOR SUBROUTINE SEQVEC*****
THE COMPLEX VECTOR V IS DENOTED
V = VMAG /_ VANGLE,WHERE VMAG IS THE MAGNITUDE
AND VANGLE IS THE ANGLE(IN DEGREES) OF V.
IN OTHER WORDS, ENTER V IN POLAR COORDINATES.

ENTER THE FIRST ELEMENT OF THE VECTOR  VPMAG
? 200
ENTER THE SECOND ELEMENT OF THE VECTOR  VPMAG
? 190
ENTER THE THIRD ELEMENT OF THE VECTOR  VPMAG
? 215
THE COMPLEX VECTOR V IS DENOTED
V = VMAG /_ VANGLE,WHERE VMAG IS THE MAGNITUDE
AND VANGLE IS THE ANGLE(IN DEGREES) OF V.
IN OTHER WORDS, ENTER V IN POLAR COORDINATES.

ENTER THE FIRST ELEMENT OF THE VECTOR  VPANGLE
? 0
ENTER THE SECOND ELEMENT OF THE VECTOR  VPANGLE
? 250
ENTER THE THIRD ELEMENT OF THE VECTOR  VPANGLE
? 110
USE THE Cntrl PRINT SCREEN OPTION NOW IF YOU WANT TO PRINT THE

RESULTS


PRESS RETURN TO CONTINUE




THE MATRIX  VP = VPMAG /_ VPANGLE IS

 200   0 DEG
 190   250 DEG
 215   110 DEG

THE MATRIX  VS = VSMAG /_ VSANGLE IS

 21.93908  20.91195 DEG
 199.621  -.4153456 DEG
 21.09858  197.6118 DEG


WHICH SUBROUTINE(ENTER AN INTEGER FROM 1 TO 4)? 4
```

4. INTENTIONALLY LEFT BLANK

(There is no software for this Chapter.)

5. LINE CONSTANTS (CHAPTER 5)

This program computes the series impedance and shunt admittance matrices of overhead single- and double-circuit three-phase transmission lines. The program also computes the electric field strength at the surface of the phase conductors and a lateral profile of ground-level electric field strength.

When the program is run, the user selects one of the following: load the input data, update the input data, save the input data, run the program, display the output, and exit.

The user can store the input data for up to five separate cases in data files called LC1.DAT, LC2.DAT, LC3.DAT, LC4.DAT, and LC5,DAT. When load or save is selected, the user is asked to enter an integer from one to five. The program then loads the input data from or saves the input data to the input data file that is selected.

When the user selects update, the input data is displayed on the monitor (for example, see the enclosed sample run). The top row of the input data consists of: the number of circuits NC, where NC=1 for single-circuit or 2 for double-circuit; earth resistivity RHO, Ωm; rated line voltage, kV line-to-line; frequency F, Hz; right-of-way width ROW, m; and the number of neutral wires N1 for circuit 1 and N2 for circuit 2, where N1 and N2 can be 0, 1, or 2.

After the first row of data, there is a row of data for each phase and for each neutral, as follows: conductor resistance R, Ω/km; conductor GMR, cm; conductor outside diameter D, cm; horizontal position X from the center of the right-of-way and vertical position Y, m; number of bundled conductors Nb, where Nb=1, 2, 3 or 4; and the bundle spacing d, cm.

When NC = 1 for a single-circuit, the input data for circuit 2 are not used when the program is run. Also, when N1 or N2 = 0 for no neutral wires, the input data for NEUTRAL 1 and NEUTRAL 2 are not used when the program is run. Similarly, when N1 or N2 = 1, the input data for NEUTRAL 2 are not used.

To make a data change, select update. The program will then display the input data and ask the use "are these correct (Y/N)?" The user should respond by pressing N for no. Next press either the up, down, left, or right cursor on the keyboard until the cursor is positioned under the number to be changed. After all changes are made, press the ESC key. The user can then save this data, run the program, and display the outputs.

After selecting run and the program has completed its computations, select display to obtain the output data. The output data consist of: the series phase impedance matrix Z_p, series sequence impedance matrix Z_s, shunt sequence impedance matrix Y_p, shunt sequence admittance matrix Y_s, conductor surface electric field strength, and a plot of the ground-level electric field strength from the center to the edge of the right-of-way. The output also includes impedance and admittance matrices for completely transposed lines.

The following sample run is for a 765-kV, 60-Hz, single-circuit three-phase line having four ACSR 954 kcmil conductors per bundle, with 45.7 cm between conductors in the bundle. The line has flat horizontal phase spacing with 13.7 m between adjacent phases and 23 m average line height. There are two Alumoweld 7 no. 8 neutral wires located 10.5 m vertically above and ±11 m horizontally from the center phase. The earth resistivity is 100 Ω m and the right-of-way width is 120 m. This sample run is also given in Example 5.10 of the text *Power System Analysis and Design with Personal Computer Applications*.

SAMPLE RUN CHAPTER 5 765-kV SINGLE-CIRCUIT

```
êêêêêêêêêêêêêêêêêêêêêêêêêêêêêêêêêêêêêêêêêêêêêêêêêêêêêêêêêêêêêêêêêêêêêêêêêêêêêê£
¤                                                                          ¤
¤                              MAIN MENU                                   ¤
äêêêêêêêêêêêêêêêêêêêêêêêêêêêêêêêêêêêêêêêêêêêêêêêêêêêêêêêêêêêêêêêêêêêêêêêêêêêêêê¿
¤                                                                          ¤
¤                                                                          ¤
¤                  L <oad input data                                       ¤
¤                  U <pdate input data                                     ¤
¤                  S <ave input data from Disk                             ¤
¤                  R <un the program                                       ¤
¤                  D <isplay the output                                    ¤
¤                  E <xit                                                  ¤
¤                                                                          ¤
¤                                                                          ¤
¤                                                                          ¤
¤                  Enter Function: <L>                                     ¤
¤                                   _                                      ¤
äêêêêêêêêêêêêêêêêêêêêêêêêêêêêêêêêêêêêêêêêêêêêêêêêêêêêêêêêêêêêêêêêêêêêêêêêêêêêêê¥
```

```
INPUT  DATA  TABLE  LC1.DAT
---------------------------------------------------------------------------
NC =      RHO =      V =      F =    ROW =    N1 =      N2 =
---------------------------------------------------------------------------
      R        GMR       D       X      Y      Nb        d
---------------------------------------------------------------------------

CIRCUIT 1
PHASE a
PHASE b
PHASE c
NEUTRAL 1
NEUTRAL 2
---------------------------------------------------------------------------

CIRCUIT 2
PHASE a
PHASE b
PHASE c
NEUTRAL 1
NEUTRAL 2
---------------------------------------------------------------------------
INPUT DATA TABLE # FROM 1 TO 5 [LC1.DAT]? 1
```

15

```
êëëëëëëëëëëëëëëëëëëëëëëëëëëëëëëëëëëëëëëëëëëëëëëëëëëëëëëëëëëëëëëëëëëëëëëëëëëëëëëëëëëë¡¡
¤                                                                                ¤
¤                              MAIN MENU                                          ¤
¤                                                                                ¤
äëëëëëëëëëëëëëëëëëëëëëëëëëëëëëëëëëëëëëëëëëëëëëëëëëëëëëëëëëëëëëëëëëëëëëëëëëëëëëëëëëëë¿
¤                                                                                ¤
¤                                                                                ¤
¤                                                                                ¤
¤              L <oad input data                                                 ¤
¤              U <pdate input data                                               ¤
¤              S <ave input data from Disk                                       ¤
¤              R <un the program                                                 ¤
¤              D <isplay the output                                              ¤
¤              E <xit                                                            ¤
¤                                                                                ¤
¤                                                                                ¤
¤                                                                                ¤
¤              Enter Function: <U>                                               ¤
¤                                                                                ¤
âëëëëëëëëëëëëëëëëëëëëëëëëëëëëëëëëëëëëëëëëëëëëëëëëëëëëëëëëëëëëëëëëëëëëëëëëëëëëëëëëëëë¥
```

$$\text{INPUT \quad DATA \quad TABLE \quad LC1.DAT}$$

	R	GMR	D	X	Y	Nb	d
NC = 1	RHO = 100	V = 765	F = 60	ROW = 120	N1 = 2	N2 = 2	
CIRCUIT 1							
PHASE a	.0701	1.229	3.038	-13.7	23	4	45.7
PHASE b	.0701	1.229	3.038	0	23	4	45.7
PHASE c	.0701	1.229	3.038	13.7	23	4	45.7
NEUTRAL 1	1.52	.0636	.978	-11	33.5	1	0
NEUTRAL 2	1.52	.0636	.978	11	33.5	1	0
CIRCUIT 2							
PHASE a	0	0	0	0	0	0	0
PHASE b	0	0	0	0	0	0	0
PHASE c	0	0	0	0	0	0	0
NEUTRAL 1	0	0	0	0	0	0	0
NEUTRAL 2	0	0	0	0	0	0	0

```
                  ARE THESE CORRECT (Y/N) ? Y
```

```
êëëëëëëëëëëëëëëëëëëëëëëëëëëëëëëëëëëëëëëëëëëëëëëëëëëëëëëëëëëëëëëëëëëëëëëëëëëëëëëëëëëë¡
¤                                                                                ¤
¤                              MAIN MENU                                          ¤
¤                                                                                ¤
äëëëëëëëëëëëëëëëëëëëëëëëëëëëëëëëëëëëëëëëëëëëëëëëëëëëëëëëëëëëëëëëëëëëëëëëëëëëëëëëëëëë¿
¤                                                                                ¤
¤                                                                                ¤
¤                                                                                ¤
¤              L <oad input data                                                 ¤
¤              U <pdate input data                                               ¤
¤              S <ave input data from Disk                                       ¤
¤              R <un the program                                                 ¤
¤              D <isplay the output                                              ¤
¤              E <xit                                                            ¤
¤                                                                                ¤
¤                                                                                ¤
¤                                                                                ¤
¤              Enter Function: <R>                                               ¤
¤                                                                                ¤
âëëëëëëëëëëëëëëëëëëëëëëëëëëëëëëëëëëëëëëëëëëëëëëëëëëëëëëëëëëëëëëëëëëëëëëëëëëëëëëëëëëë¥
```

```
ëëëëëëëëëëëëëëëëëëëëëëëëëëëëëëëëëëëëëëëëëëëëëëëëëëëëëëëëëëëëëëëëëëëëëëëëë£
¤                                                                      ¤
¤                          MAIN MENU                                   ¤
¤                                                                      ¤
äëëëëëëëëëëëëëëëëëëëëëëëëëëëëëëëëëëëëëëëëëëëëëëëëëëëëëëëëëëëëëëëëëëëëëëëë¿
¤                                                                      ¤
¤                                                                      ¤
¤                                                                      ¤
¤              L <oad input data                                       ¤
¤              U <pdate input data                                     ¤
¤              S <ave input data from Disk                             ¤
¤              R <un the program                                       ¤
¤              D <isplay the output                                    ¤
¤              E <xit                                                  ¤
¤                                                                      ¤
¤                                                                      ¤
¤                                                                      ¤
¤            Enter Function: <D>                                       ¤
¤                                                                      ¤
àëëëëëëëëëëëëëëëëëëëëëëëëëëëëëëëëëëëëëëëëëëëëëëëëëëëëëëëëëëëëëëëëëëëëëëëë¥
```

            ENTER YOUR CASE NAME SAMPLE RUN

```
ëëëëëëëëëëëëëëëëëëëëëëëëëëëëëëëëëëëëëëëëëëëëëëëëëëëëëëëëëëëëëëëëëëëëëëëëë£
¤                                                                      ¤
¤                                                                      ¤
¤    SINGLE CIRCUIT                    SAMPLE RUN                       ¤
¤                                                                      ¤
¤    SERIES PHASE IMPEDANCE MATRIX      Zp   EQ. 5-7-19       ohms/km   ¤
¤                                                                      ¤
äëëëëëëëëëëëëëëëëëëëëëëëëëëëëëëëëëëëëëëëëëëëëëëëëëëëëëëëëëëëëëëëëëëëëëëëë¿
¤                                                                      ¤
¤ 0.1181E+00 + j 5.532E-01 0.1009E+00 + j 2.340E-01 0.9813E-01 + j 1.842E-01 ¤
¤                                                                      ¤
¤ 0.1009E+00 + j 2.339E-01 0.1200E+00 + j 5.500E-01 0.1009E+00 + j 2.339E-01 ¤
¤                                                                      ¤
¤ 0.9813E-01 + j 1.842E-01 0.1009E+00 + j 2.340E-01 0.1181E+00 + j 5.532E-01 ¤
¤                                                                      ¤
àëëëëëëëëëëëëëëëëëëëëëëëëëëëëëëëëëëëëëëëëëëëëëëëëëëëëëëëëëëëëëëëëëëëëëëëë¥
```

```
ëëëëëëëëëëëëëëëëëëëëëëëëëëëëëëëëëëëëëëëëëëëëëëëëëëëëëëëëëëëëëëëëëëëëëëëëë£
¤                                                                      ¤
¤                                                                      ¤
¤    SINGLE CIRCUIT                    SAMPLE RUN                       ¤
¤                                                                      ¤
¤    Zp FOR A COMPLETELY TRANSPOSED LINE   Zphat  EQ. 5-7-21  ohms/km   ¤
¤                                                                      ¤
äëëëëëëëëëëëëëëëëëëëëëëëëëëëëëëëëëëëëëëëëëëëëëëëëëëëëëëëëëëëëëëëëëëëëëëëë¿
¤ 0.1187E+00 + j 5.522E-01 0.9997E-01 + j 2.174E-01 0.9997E-01 + j 2.174E-01 ¤
¤                                                                      ¤
¤ 0.9997E-01 + j 2.174E-01 0.1187E+00 + j 5.522E-01 0.9997E-01 + j 2.174E-01 ¤
¤                                                                      ¤
¤ 0.9997E-01 + j 2.174E-01 0.9997E-01 + j 2.174E-01 0.1187E+00 + j 5.522E-01 ¤
¤                                                                      ¤
àëëëëëëëëëëëëëëëëëëëëëëëëëëëëëëëëëëëëëëëëëëëëëëëëëëëëëëëëëëëëëëëëëëëëëëëë¥
```

```
SINGLE CIRCUIT                          SAMPLE RUN
  SERIES SEQUENCE IMPEDANCE MATRIX Zs      EQ. 5-7-25        ohms/km
```

```
 0.3187E+00 + j 9.869E-01  0.1264E-01 - j 9.112E-03  -.1421E-01 - j 6.389E-03
 -.1421E-01 - j 6.374E-03  0.1875E-01 + j 3.348E-01  -.2903E-01 + j 1.814E-02
 0.1262E-01 - j 9.117E-03  0.3022E-01 + j 1.607E-02  0.1875E-01 + j 3.348E-01
```

```
  SINGLE CIRCUIT                            SAMPLE RUN
   Zs FOR A COMPLETELY TRANSPOSED LINE    Zshat  EQ. 5-7-27   ohms/km
```

```
 0.3187E+00 + j 9.869E-01  0.0000E+00 + j 0.000E+00  0.0000E+00 + j 0.000E+00
 0.0000E+00 + j 0.000E+00  0.1875E-01 + j 3.348E-01  0.0000E+00 + j 0.000E+00
 0.0000E+00 + j 0.000E+00  0.0000E+00 + j 0.000E+00  0.1875E-01 + j 3.348E-01
```

```
  SINGLE CIRCUIT                            SAMPLE RUN
   SHUNT PHASE ADMITTANCE MATRIX    Yp   EQ. 5-11-16        s/km
```

```
 + j 4.311E-06   - j 7.666E-07   - j 2.167E-07
 - j 7.666E-07   + j 4.439E-06   - j 7.666E-07
 - j 2.167E-07   - j 7.666E-07   + j 4.311E-06
```

```
  SINGLE CIRCUIT                            SAMPLE RUN
   Yp FOR A COMPLETELY TRANSPOSED LINE   Yphat EQ. 5-11-17  s/km
```

```
 + j 4.354E-06   - j 5.833E-07   - j 5.833E-07
 - j 5.833E-07   + j 4.354E-06   - j 5.833E-07
 - j 5.833E-07   - j 5.833E-07   + j 4.354E-06
```

```
èëëëëëëëëëëëëëëëëëëëëëëëëëëëëëëëëëëëëëëëëëëëëëëëëëëëëëëëëëëëëëëëëëëëëëëëëë£
¤                                                                        ¤
¤     SINGLE CIRCUIT                      SAMPLE RUN                      ¤
¤                                                                        ¤
¤     SHUNT SEQUENCE ADMITTANCE MATRIX  Ys    EQ. 5-11-20      s/km       ¤
¤                                                                        ¤
äëëëëëëëëëëëëëëëëëëëëëëëëëëëëëëëëëëëëëëÜëëëëëëëëëëëëëëëëëëëëëëëëëëëëëëëëëëëë¿
¤  0.0000E+00 + j 3.187E-06 -.1219E-06 + j 7.036E-08 0.1219E-06 + j 7.036E-08  ¤
¤                                                                        ¤
¤  0.1219E-06 + j 7.036E-08 -.3901E-13 + j 4.937E-06 0.3544E-06 - j 2.046E-07  ¤
¤                                                                        ¤
¤ -.1219E-06 + j 7.036E-08 -.3544E-06 - j 2.046E-07 0.3901E-13 + j 4.937E-06  ¤
¤                                                                        ¤
àëëëëëëëëëëëëëëëëëëëëëëëëëëëëëëëëëëëëëëëëëëëëëëëëëëëëëëëëëëëëëëëëëëëëëëëëëë¥
```

```
èëëëëëëëëëëëëëëëëëëëëëëëëëëëëëëëëëëëëëëëëëëëëëëëëëëëëëëëëëëëëëëëëëëëëëëëëë£
¤                                                                        ¤
¤     SINGLE CIRCUIT                      SAMPLE RUN                      ¤
¤                                                                        ¤
¤     Ys FOR A COMPLETELY TRANSPOSED LINE    Yshat EQ. 5-11-23   s/km     ¤
¤                                                                        ¤
äëëëëëëëëëëëëëëëëëëëëëëëëëëëëëëëëëëëëëëëëëëëëëëëëëëëëëëëëëëëëëëëëëëëëëëëëëë¿
¤  0.0000E+00 + j 3.187E-06 0.0000E+00 + j 0.000E+00 0.0000E+00 + j 0.000E+00  ¤
¤                                                                        ¤
¤  0.0000E+00 + j 0.000E+00 0.0000E+00 + j 4.937E-06 0.0000E+00 + j 0.000E+00  ¤
¤                                                                        ¤
¤  0.0000E+00 + j 0.000E+00 0.0000E+00 + j 0.000E+00 0.0000E+00 + j 4.937E-06  ¤
¤                                                                        ¤
àëëëëëëëëëëëëëëëëëëëëëëëëëëëëëëëëëëëëëëëëëëëëëëëëëëëëëëëëëëëëëëëëëëëëëëëëëë¥
```

CONDUCTOR SURFACE ELECTRIC FIELD STRENGTH Eqs.5-12-1 to 5-12-5

        PHASE: 1 . 17.93178   kV/cm
        PHASE: 2 . 19.34038   kV/cm
        PHASE: 3 . 17.93178   kV/cm

     MAXIMUM ELECTRIC FIELD IS: 19.34038 kV/cm

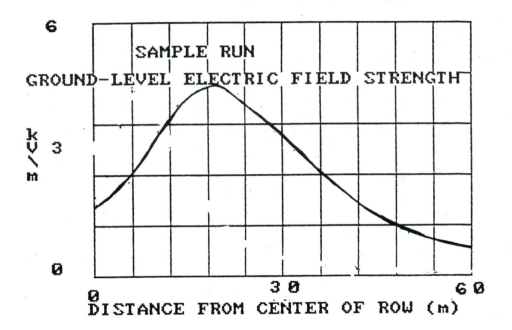

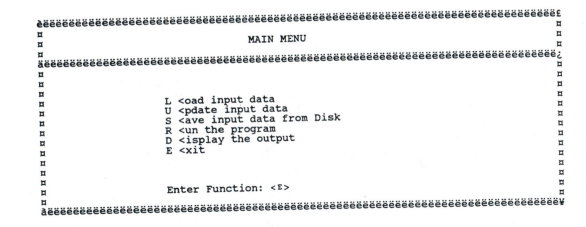

6. TRANSMISSION LINES-STEADY STATE OPERATION (CHAPTER 6)

This program computes the *ABCD* parameters and equivalent π circuit values for a single-phase or completely transposed three-phase transmission line. The program also computes the sending end quantities: voltage; current; real, reactive, and apparent power; and power factor for a given receiving-end load, with or without compensation, assuming balanced steady-state operation. Voltage regulation and maximum power flow are also computed.

Input data to the program consist of: (1) rated line voltage and line length; (2) R, X_L, G, and B_C; (3) receiving-end full-load voltage, apparent power, and power factor; (4) the number (0, 1, or 2) and location of intermediate substations; and (5) the percent shunt reactive (inductive) and series capacitive compensation at each line terminal and at each intermediate substation. It is assumed that the shunt reactors are removed at heavy loads.

The output data consist of (a) without compensation: (1) characteristic impedance, (2) propagation constant; (3) wavelength, (4) surge impedance loading, (5) equivalent π circuit series impedance and shunt admittance, and (6) *ABCD* parameters; and (b) with compensation: (1) equivalent *ABCD* parameters, (2) sending-end voltage, current, real, reactive, and apparent power, and power factor; (3) percent voltage regulation; and (4) theoretical maximum real power delivered to the receiving end with rated terminal voltages.

The following sample run is for a three-phase 345-kV, 200-km, 60 Hz line with positive sequence impedance 0.047 + j 0.37 Ω/km and positive sequence shunt admittance 0 + j 4.1E-6 S/km. Full-load at the receiving-end of the line is 800 MVA at 0.99 power factor leading and at 328 kV. 75% total shunt reactive compensation is installed, half (37.5%) at each end of the line. The shunt reactors are removed at heavy loads. There are no intermediate substations and there is no series compensation.

SAMPLE RUN CHAPTER 6 345-kV 200-km LINE

```
THIS PROGRAM EVALUATES A SINGLE PHASE OR
A BALANCED THREE PHASE TRANSMISSION LINE
UNDER STEADY STATE OPERATING CONDITIONS

THE FOLLOWING QUANTITIES ARE COMPUTED
(A)WITHOUT COMPENSATION
    (1)CHARACTERISTIC IMPEDANCE, EQ(6.2.16)
    (2)PROPAGATION CONSTANT, EQ(6.2.12)
    (3)WAVELENGTH, EQ(6.4.15)
    (4)SURGE IMPEDANCE LOADING, EQ(6.4.21)
    (5)EQUIVALENT PI CIRCUIT SERIES IMPEDANCE, EQ(6.3.5)-EQ(6.3.6)
    (6)EQUIVALENT PI CIRCUIT SHUNT ADMITTANCE, EQ(6.3.1)-EQ(6.3.3)
    (7)ABCD PARAMETERS, EQ(6.3.1)-EQ(6.3.3)
(B)WITH COMPENSATION
    (1)EQUIVALENT FULL-LOAD ABCD PARAMETERS
    (2)SENDING END VOLTAGE,CURRENT AND POWER, EQ(6.1.3)
    (3)% VOLTAGE REGULATION, EQ(6.1.8)
    (4)THEORETICAL MAXIMUM REAL POWER DELIVERED
        TO THE RECEIVING END WITH RATED TERMINAL VOLTAGES
```

```
LINE DATA

ENTER THE NAME OF THE LINE(ANY ALPHANUMERIC SEQUENCE) SAMPLE RUN
IS THIS A THREE PHASE LINE(ENTER Y)OR A SINGLE PHASE LINE(ENTER N) ? Y
ENTER THE RATED LINE VOLTAGE(LINE-LINE kV) 345
ENTER THE LINE LENGTH(km) 200
ENTER THE SERIES RESISTANCE(ohms/km) 0.047
ENTER THE SERIES REACTANCE(ohms/km) 0.37
ENTER THE SHUNT CONDUCTANCE(mhos/km) C
ENTER THE SHUNT SUSCEPTANCE(mhos/km) 4.1E-6


RECEIVING END DATA

ENTER THE RECEIVING END FULL-LOAD VOLTAGE(LINE-LINE kV) 328
ENTER THE RECEIVING END FULL-LOAD APPARENT POWER(MVA) 800
ENTER THE RECEIVING END FULL-LOAD POWER FACTOR(PER UNIT) 0.99
IS THE POWER FACTOR LAGGING(ENTER Y)OR LEADING(ENTER N) ? N


IS THE LINE COMPENSATED(Y OR N) ? Y

COMPENSATION INPUT DATA AT EACH LOCATION
IS GIVEN IN PER CENT OF TOTAL SHUNT SUSCEPTANCE
OR SERIES REACTANCE OF THE LINE


ENTER THE % SHUNT REACTIVE COMPENSATION AT THE SENDING END 37.5
ENTER THE % SERIES CAPACITIVE COMPENSATION AT THE SENDING END 0
ENTER THE % SHUNT REACTIVE COMPENSATION AT THE RECEIVING END 37.5
ENTER THE % SERIES CAPACITIVE COMPENSATION AT THE RECEIVING END 0
ENTER THE NUMBER OF INTERMEDIATE SUBSTATIONS(0,1,OR 2): 0

DO YOU WANT TO CONTINUE (Y) OR RESET THE INPUT DATA (N) ? Y



USE THE Ctrl PRINT SCREEN OPTION NOW IF YOU WANT TO PRINT THE RESULTS
PRESS RETURN TO CONTINUE
**********************INPUT DATA*************************
SAMPLE RUN      THREE-PHASE    345 kV     200 km
  LINE DATA:
SERIES RESISTANCE (ohms/km)                4.700E-02
SERIES REACTANCE (ohms/km)                 3.700E-01
SHUNT CONDUCTANCE (mhos/km)                0.000E+00
SHUNT SUSCEPTANCE (mhos/km)                4.100E-06

RECEIVING END DATA:
RECEIVING END FULL-LOAD VOLTAGE (kV)       3.280E+02
RECEIVING END FULL-LOAD
APPARENT POWER (MVA)                       8.000E+02
RECEIVING END FULL-LOAD
POWER FACTOR (per unit)                    9.900E-01LEADING

COMPENSATION DATA:
SHUNT COMPENSATION AT SENDING END(%)       37.5
SERIES COMPENSATION AT SENDING END(%)      0
SHUNT COMPENSATION AT RECEIVING END(%)     37.5
SERIES COMPENSATION AT RECEIVING END(%)    0
NUMBER OF INTERMEDIATE SUBSTATIONS         0
TOTAL SHUNT COMPENSATION(%)                75
TOTAL SERIES COMPENSATION(%)               0
```

```
**********************OUTPUT DATA*************************

    *****WITHOUT COMPENSATION*****
CHARACTERISTIC IMPEDANCE(ohms)              3.016E+02/ -3.620E+00DEG
PROPAGATION CONSTANT(1/m)                   7.807E-05 + j 1.234E-03
WAVELENGTH(km)                              5.091E+03
SURGE IMPEDANCE LOADING(MW)                 3.946E+02

EQUIVALENT PI CIRCUIT:
SERIES IMPEDANCE(ohms)                      9.211E+00  +j 7.327E+01
SHUNT ADMITTANCE(mhos)                      5.336E-07  +j 8.242E-04
ABCD PARAMETERS:
A PARAMETER(per unit)                       9.698E-01/  2.254E-01DEG
B PARAMETER(ohms)                           7.384E+01/  8.283E+01DEG
C PARAMETER(mhos)                           8.117E-04/  9.007E+01DEG
D PARAMETER(per unit)                       9.698E-01/  2.254E-01DEG

    ******WITH COMPENSATION******
FULL-LOAD ABCD PARAMETERS
(SHUNT COMPENSATION IS REMOVED):
A PARAMETER (per unit)                      9.698E-01/  2.254E-01DEG
B PARAMETER (ohms)                          7.384E+01/  8.283E+01DEG
C PARAMETER (mhos)                          8.117E-04/  9.007E+01DEG
D PARAMETER (per unit)                      9.698E-01/  2.254E-01DEG

NO-LOAD ABCD PARAMETERS
(SHUNT AND SERIES COMPENSATION ARE INCLUDED):
A PARAMETER (per unit)                      9.925E-01/  5.591E-02DEG
B PARAMETER(ohms)                           7.384E+01/  8.283E+01DEG
C PARAMETER(mhos)                           2.053E-04/  8.988E+01DEG
D PARAMETR (per unit)                       9.925E-01/  5.591E-02DEG

SENDING END VOLTAGE(LINE-TO-LINE kV)        3.636E+02/  2.992E+01 (+30 )
SENDING END CURRENT(kA)                     1.396E+00/  1.459E+01DEG
SENDING END REAL POWER(MW)                  8.479E+02
SENDING END REACTIVE POWER(Mvars)           2.324E+02
SENDING END APPARENT POWER(MVA)             8.792E+02
SENDING END POWER FACTOR(per unit)          9.644E-01 LAGGING

VOLTAGE REGULATION(%)                       1.169E+01
THEORETICAL MAXIMUM REAL POWER
DELIVERED(MW)                               1.411E+03


REMOVE Ctrl PRINT SCREEN AND PRESS RETURN TO CONTINUE.
```

7. POWER FLOW (CHAPTER 7)

This application computes the voltage magnitude and phase angle at each bus in a power system under balanced three-phase steady-state operation. Bus voltages and phase angles are then used to compute generator, line, and transformer loadings.

The maximum number of buses is 100. Also, the maximum number of transmission lines or transformers is 100.

Input data include bus data, transmission line data, and transformer data, which are stored in three data files. The user can store the input data for any number of separate cases. The bus, line, and transformer input data files have the file extensions *.PFB, *.PFL, and *.PFT respectively.

APPLICATION OPERATION

The application is started in the usual Windows manner by double clicking on the "Chapter 7 Power Flow" icon in the **Program Manager** "Power System Analysis and Design" window.

The application presents the user with a main parent window entitled "Chapter 7 Power Flow" holding three smaller child windows, containing the input data forms. The child windows are entitled "Bus Data", ."Line Data", "Transformer Data". See Figure 7a.

The Title Bar at the top of parent window has the usual Windows Control Box button on the left hand side, and the Minimize and Maximize buttons on the right hand side. The Menu bar at the top of the parent window is described later in this chapter. The bottom of the parent window shows a Status bar. The Status bar shows the active child window and the value of the current cell in that child window.

ENTERING DATA

Select either the Bus Data, Line Data or Transformer Data form to make it the active form. Selection can be achieved by either clicking on the form, or selecting the appropriate item in the **Window** menu. The cell for the data to be entered into is selected by either clicking on the cell, or using the arrow keys to move around the cells. The contents of the active cell is displayed in the status bar. The contents of the cell can be erased by using either the delete key to remove the entire contents, or the backspace key to remove the left hand character in the string. Values can be entered in the form:

> Real
> 123.45
> 1.2345e2

Bus input data include: (1) bus number, automatically inserted in the form by the application, (2) bus type (0 for the swing bus, 1 for a load bus, 2 for a voltage-controlled bus, and 3 for a voltage-controlled bus with a tap changing transformer), (3) per-unit voltage magnitude for the swing bus and for the voltage-controlled buses, (4) phase angle (degrees) for the swing bus, (5) per-unit real generation, (6) per-unit reactive generation, (7) per-unit real load, (8) per-unit reactive load, and (9) maximum and minimum limits on per-unit reactive generative generation for voltage-controlled (type 2) buses. The swing bus is always bus 1 for this application.

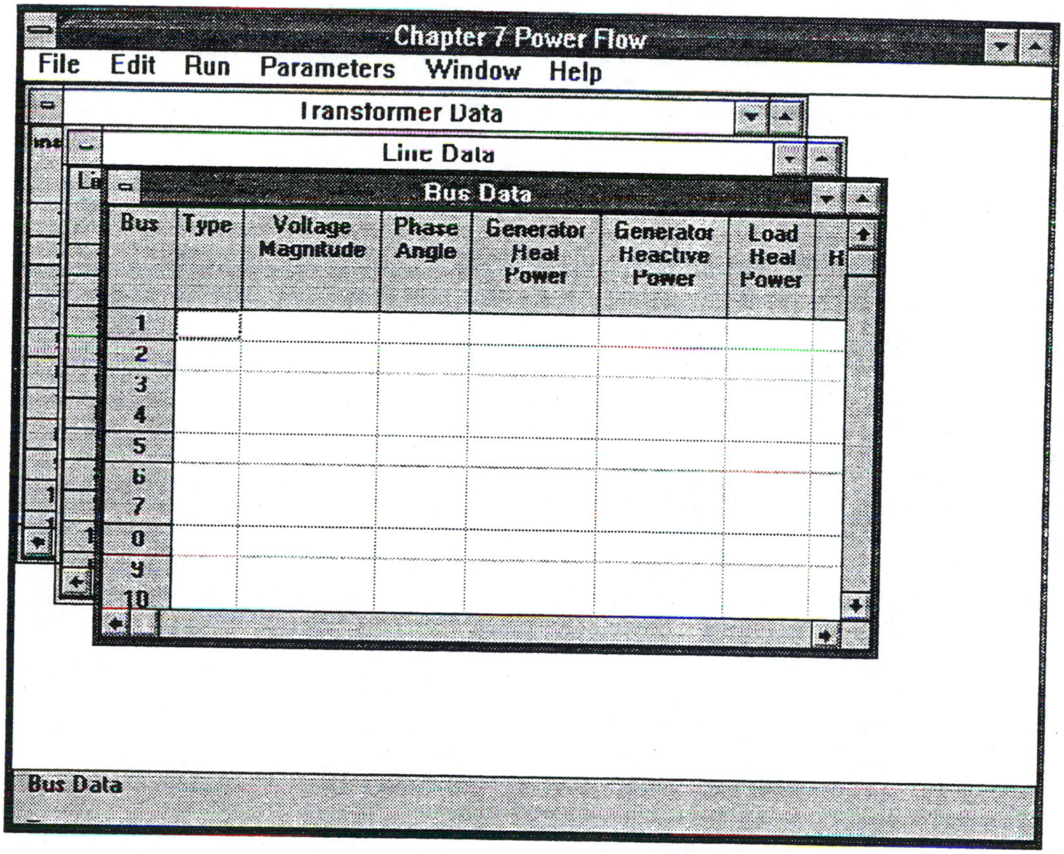

Figure 7a. Power Flow Parent Window

25

Line input data include: (1) line number, automatically inserted in the form by the application, (2) the two buses to which the line is connected, (3) per-unit series resistance, (4) per-unit series reactance, (5) per-unit shunt conductance, (6) per-unit shunt susceptance, and (7) per-unit maximum line loading (apparent power).

Transformer data include: (1) transformer number, automatically inserted in the form by the application, (2) the two buses to which the transformer is connected, (3) per-unit winding resistance, (4) per-unit leakage reactance, (5) per-unit core-loss conductance, (6) per-unit magnetizing susceptance, (7) per-unit maximum transformer loading (apparent power), and (8) per-unit tap setting (values between 0.75 to 1.25, see following notes).

If the transformer shunt branch is neglected, enter zero values for (5) core-loss conductance and (6) magnetizing susceptance. If the shunt branch is included, enter a positive value for (5) per-unit core-loss conductance, and a <u>negative</u> value for (6) per-unit magnetizing susceptance. The program places one-half of the shunt branch on each side of the leakage impedance.

The tap setting in this program is the reciprocal of the transformer turns ratio c shown in Figure 4.27 of the text.

TYPE 3 BUS

A voltage-controlled bus with a tap-changing transformer is designated when the user enters bus type 3 in the bus input data. For type 3 buses, there must be a transformer in the transformer input data file whose <u>second</u> bus number is the voltage-controlled bus. The application will warn the user when there is a type 3 bus and no transformer whose second bus number is a type 3 bus. Also, the application assumes that the tap is on the same side of the transformer as the <u>second</u> bus number of that transformer.

The application varies the tap setting of each tap-changing transformer connected to a type 3 bus until either: (1) the computed bus voltage is within 0.4% of the input bus voltage, or (2) the maximum or minimum tap setting is reached. The user inputs a value from 1.00 to 1.25 for the maximum tap setting MAXTAP of each transformer in the transformer input data form, Tap Setting column. The application sets the minimum tap setting to MINTAP=(2.0-MAXTAP).

FIXED TRANSFORMER TAP SETTING

The user can also set tap settings of tap-changing transformers to fixed values. The user enters the fixed-tap setting (values from 0.75 to 1.25) into the per-unit tap setting column of the transformer input data form.

STARTING VALUES

The user has the following options for starting values: (1) **Flat Start**, or (2) **Hot Start**, using output values from the previous run. If the user selects a **Flat Start**, then the initial voltage magnitudes of type 1 buses are set equal to the swing bus voltage, and all initial phase angles are set equal to the swing bus phase angle. Also, the initial tap setting of all the variable tap-changing transformers are set equal to 1.0. If the user selects a **Hot Start**, then the initial bus voltage magnitude, phase angle, and transformer tap setting values used are taken from the previous power flow results.

RUNNING POWER FLOW AND OUTPUT DATA

When preparing a "base case" for the first time with no output data, then select a **Flat Start**. After running for a base case with a successful power-flow solution, the application automatically updates the voltage magnitude, phase angle and tap setting values in the bus input data form. Selecting a **Hot Start** when you make a change such as removing a line or transformer will cause these values to be used. The application will usually converge more rapidly when you start with the output values from a previous (and successful) run.

Output results include bus output, line output, and transformer output. The application gives the user the option of displaying the results on the monitor and on the printer.

Bus output results include: (1) bus number, (2) per-unit voltage magnitude, (3) voltage phase angle (degrees), (4) per-unit real generation, (5) per-unit reactive generation, (6) per-unit real load power, and (7) per-unit reactive load power. Those buses for which the voltage magnitude is 5% higher or lower than the swing bus voltage are identified.

Line output results include: (1) line number, (2) the two buses to which the line is connected, (3) the per-unit real, reactive and apparent power flows into each end of the line. Those lines whose loadings exceed their maximum loadings are also identified.

Transformer output results include: (1) transformer number, (2) the two buses to which the transformer is connected, (3) the per-unit real, reactive, and apparent power flows into each winding, and (4) per-unit tap setting. Those transformers whose loadings exceed their maximum loadings are identified.

MENUS

The Menu bar has the following groups: **File, Edit, Run, Parameters, Options, Window, Help**.

The **File** menu has the following items:

Item	Action	Short Cut Key
New Study	Clears all three input data forms, Bus Data, Line Data, and Transformer Data.	
Open Study	Opens the three files to read in the bus, line and transformer data into the appropriate input data forms. The user is prompted for each filename in turn. However the default filename is taken to be that given for the bus data. But the user can select different filenames for the line and transformer data files if necessary.	
Save Study	Saves all input data form values for the bus, line and transformer data in three files.	
Save Study As	Saves all input data form values for the bus, line and transformer data in three files, allowing the user to specify a filename before saving.	
New File	Clears the active input data form.	
Open File	Opens the appropriate file to read in the data into active input data form.	
Save File	Saves the active input data form values in a file.	
Save File As	Saves the active input data form values in a file, allowing the user to specify a filename before saving.	
Print	Prints the data or results from the active child window of input data form on the default Windows printer.	Ctrl+P
Transient Stability	This item starts a dialog that prepares the data files necessary to execute the Chapter 13 TRANSIENT STABILITY program. Data is taken from the last power flow solution calculated. The data is stored in one of three files named either TSPF1.DAT, TSPF2.DAT or TSPF3.DAT. These files correspond to cases 1, 2 or 3 respectively, in the Chapter 13 program menu.	
Exit	Exits the application, saving any input data as required.	

The **Edit** menu has the following items:

Item	Action	Short Cut Key
Undo	Undo last edit operation on active input data form.	Ctrl+Z
Cut	Cut contents of selected cells in active input data form or child window to the Clipboard.	Ctrl+X
Copy	Copy contents of selected cells in active input data form or child window to the Clipboard.	Ctrl+C
Paste	Paste contents of Clipboard into selected cells in active input data form.	Ctrl+V
Delete	Delete contents of selected cells in active input data form.	Del
Select All	Select all cells in the active input data form.	

The **Run** menu has the following items:

Item	Action
Power Flow	Runs a power flow on the system entered into the input data forms.

The **Options** menu has the following items:

Item	Action
Printer Fonts	Allows the user to select printer font type, size, and style. The default is True Type Courier New, size 10, normal style. Best results are achieved with fixed space fonts, not proportional fonts.

The **Parameters** menu has the following items:

Item	Action
Tolerance	A dialog allows the user to alter the convergence tolerance for the power flow calculations.
Iterations	A dialog allows the user to alter the number of iterations that are executed during the power flow calculations before the calculations are halted with an error warning.
Starting Voltage	A dialog allows the user to specify either a **Flat Start**, or a **Hot Start** for the power flow calculations. (See the **Starting Values** section in this chapter.)

The **Window** menu has the following items:

Item	Action
Cascade	Arranges the child windows in a cascade within the main window.
Tile Horizontally	Arranges the child windows tiled horizontally within the main window.
Tile Vertically	Arranges the child windows tiled vertically within the main window.
Restore All	Restores all child window icons to their original size.
Minimize All	Minimizes all the child windows into icons.
Arrange Icons	Tidies up the arrangement of any child window icons.
Refresh	Refreshes the display.
1, 2, 3,...	Selects the listed window as the active child window.

The **Help** menu has the following item:

Item	Action
About	Lists application authors and shows system statistics.

SAMPLE RUN CHAPTER 7 FIVE-BUS POWER SYSTEM

The following sample run is for the five-bus power system whose single-line diagram is shown in Figure 7b. Bus, line, and transformer input data are given for the sample run. The data is contained in files PSAD.PFB, PSAD.PFL, PSAD.PFT. Note that bus 5 is a type 3 bus. The program adjusts the tap setting on transformer T1.

SAMPLE BUS INPUT DATA

Bus	Type	Voltage Magnitude	Phase Angle	Generator Real Power	Generator Reactive Power	Load Real Power	Load Reactive Power	Generator Maximum Reactive Power	Generator Minimum Reactive Power
1	0	1.00	0.00			0.00	0.00		
2	1			0.00	0.00	2.00	0.20		
3	2	1.04		1.30		0.20	0.10	1.00	-0.70
4	1			0.00	0.00	0.00	0.00		
5	3	1.02		0.00	0.00	0.00	0.00		

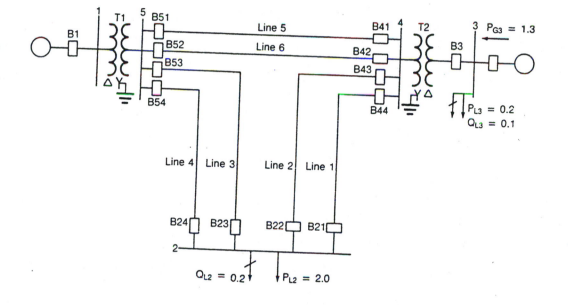

Figure 7b. Single-line diagram for Sample Run Chapter 7
(per-unit values are shown)

SAMPLE BUS INPUT DATA

Bus	Type	Voltage Magnitude	Phase Angle	Generator Real Power	Generator Reactive Power	Load Real Power	Load Reactive Power	Generator Maximum Reactive Power	Generator Minimum Reactive Power
1	0	1.00	0.00			0.00	0.00		
2	1			0.00	0.00	2.00	0.20		
3	2	1.04		1.30		0.20	0.10	1.00	-0.70
4	1			0.00	0.00	0.00	0.00		
5	3	1.02		0.00	0.00	0.00	0.00		

SAMPLE LINE INPUT DATA

Line	From Bus	To Bus	Resistance	Reactance	Conductance	Susceptance	Maximum Power Rating
1	2	4	0.036	0.40	0.00	0.43	3.00
2	2	4	0.036	0.40	0.00	0.43	3.00
3	2	5	0.018	0.20	0.00	0.22	3.00
4	2	5	0.018	0.20	0.00	0.22	3.00
5	4	5	0.009	0.10	0.00	0.11	3.00
6	4	5	0.009	0.10	0.00	0.11	3.00

SAMPLE TRANSFORMER INPUT DATA

Transformer	From Bus	To Bus	Resistance	Reactance	Conductance	Susceptance	Maximum Power Rating	Tap Setting
1	1	5	0.006	0.080	0.000	0.000	1.5	1.200
2	3	4	0.003	0.040	0.000	0.000	2.5	1.000

32

Enter the sample data into the input data forms. Figure 7c shows the entered sample data with the windows tiled horizontally. To perform the power flow calculations select the **Run** menu, **Power Flow** item. When the application has finished the calculations, two new forms will open. One entitled Analysis Messages, and the other entitled Analysis Results.

The Analysis Messages window lists the status of the calculations at every iteration. The purpose of this window is to allow the user to find the cause of the problem for any calculations that fail.

The Analysis Results window lists the results of the power flow calculation. Figure 7d shows the Analysis Results window horizontally tiled, with all the other windows minimized into icons. A complete listing of the sample run results follows:

SAMPLE RUN POWER FLOW RESULTS

POWER FLOW RESULTS   Study:
07 November 1993
16:49:55

POWER FLOW BUS OUTPUT DATA   File: SAMPLE RUN

BUS #	VOLTAGE MAGNITUDE per unit	PHASE ANGLE	GENERATION LOAD				Voltage < 0.95 > 1.05
			PG per unit	QG per unit	PL per unit	QL per unit	
1	1.0000	0.0000	0.9379	-0.8444	0.0000	0.0000	
2	1.0417	-11.3326	0.0000	0.0000	2.0000	0.2000	
3	1.0400	-1.1158	1.3000	-0.0438	0.2000	0.1000	
4	1.0432	-3.4628	0.0000	0.0000	0.0000	0.0000	
5	1.0223	-4.3135	0.0000	0.0000	0.0000	0.0000	
		TOTAL	2.2379	-0.8881	2.2000	0.3000	

MISMATCH = 6.540E-07

POWER FLOW LINE OUTPUT DATA   File: SAMPLE RUN

LINE #	BUS TO BUS		P	Q	S	RATING EXCEEDED
1	2	4	-0.3671	-0.1786	0.4082	
	4	2	0.3717	-0.2380	0.4413	
2	2	4	-0.3671	-0.1786	0.4082	
	4	2	0.3717	-0.2380	0.4413	
3	2	5	-0.6329	0.0786	0.6378	
	5	2	0.6402	-0.2319	0.6809	
5	4	5	0.1766	0.1433	0.2275	
	5	4	-0.1760	-0.2540	0.3091	
6	4	5	0.1766	0.1433	0.2275	
	5	4	-0.1760	-0.2540	0.3091	

POWER FLOW TRANSFORMER OUTPUT DATA   File: SAMPLE RUN

TRANSFORMER	BUS TO BUS		P	Q	S	TAP SETTING	RATING EXCEEDED
1	1	5	0.9379	-0.8444	1.2620	0.9600	
	5	1	-0.9283	0.9718	1.3439		
2	3	4	1.1000	-0.1438	1.1094	1.0000	
	4	3	-1.0966	0.1893	1.1128		

CUT, COPY AND PASTE

When copying between input data forms within the application, the same number and arrangement of cells must be selected, in the input data form copied from, and the input data form pasted to.

The **Cut, Copy and Paste** actions use the Windows Clipboard as the destination for **Cut** and **Copy**, and the source for **Paste**, operations. Therefore it is possible to copy the contents of any of the input data forms into another Windows application that will accept input from the Clipboard. Also values can be pasted from another application into a matrix through the Clipboard. (The values for different columns must be separated by tabs, rows are separated by carriage returns in the other application. Do not include any spaces.)

The **Edit** menu, **Select All** item selects all the cells in the active input data form. Individual cells can be selected by clicking on them. Groups of cells can be selected by clicking on the first cell, and while holding down the mouse button, dragging across the other cells in the group. An entire row (or column) can be selected by clicking on the row (or column) number (or name). Multiple rows (or columns) can be selected by clicking on the first row (or column) number (or name) and while holding down the mouse button, dragging across the other row (or column) numbers (or names) required.

File   Edit   Run   Parameters   Window   Help

Bus Data

Bus	Type	Voltage Magnitude	Phase Angle	Generator Real Power	Generator Reactive Power	Load Real Power	Load Reactive Power	Generator Maximum Reactive Power	Generator Minimum Reactive Power
1	0	1.00	0.00			0.00	0.00		
2	1			0.00	0.00	2.00	0.20		
3	2	1.04		1.30		0.20	0.10	1.00	-0.70

Line Data

Line	From Bus	To Bus	Resistance	Reactance	Conductance	Susceptance	Maximum Power Rating
1	2	4	0.036	0.40	0.00	0.43	3.00
2	2	4	0.036	0.40	0.00	0.43	3.00
3	2	5	0.018	0.20	0.00	0.22	3.00
4	2	5	0.018	0.20	0.00	0.22	3.00

Transformer Data

Transform	From Bus	To Bus	Resistance	Reactance	Conductance	Susceptance	Maximum Power Rating	Tap Setting
1	1	5	0.006	0.080	0.000	0.000	1.5	1.200
2	3	4	0.003	0.040	0.000	0.000	2.5	1.000
3								
4								

Bus Data
0

Figure 7c. Power Flow Input Data, Sample Run Chapter 7

```
╔══════════════════════════════════════════════════════════════════════════╗
║ ═                        Chapter 7 Power Flow                      ▼  ▲    ║
╠══════════════════════════════════════════════════════════════════════════╣
║  File    Window    Help                                                    ║
╠══════════════════════════════════════════════════════════════════════════╣
║  ═                        Analysis Results                      ▼  ▲      ║
║                                                                            ║
║    POWER FLOW RESULTS                    Study:                            ║
║    ------------------                    ----------------                  ║
║                                                                            ║
║    07 November 1993                                                        ║
║    16:49:55                                                                ║
║                                                                            ║
║    POWER FLOW BUS OUTPUT DATA            File:                             ║
║    -------------------------             ------------------                ║
```

| | | | GENERATION | | LOAD | | |
BUS #	VOLTAGE MAGNITUDE per unit	PHASE ANGLE degrees	PG per unit	QG per unit	PL per unit	QL per unit	Voltage < 0.95 > 1.05
1	1.0000	0.0000	0.9379	-0.8444	0.0000	0.0000	
2	1.0417	-11.3326	0.0000	0.0000	2.0000	0.2000	
3	1.0400	-1.1158	1.3000	-0.0438	0.2000	0.1000	
4	1.0432	-3.4628	0.0000	0.0000	0.0000	0.0000	
5	1.0223	-4.3135	0.0000	0.0000	0.0000	0.0000	
			--------	--------	--------	--------	
		TOTAL	2.2379	-0.8881	2.2000	0.3000	

```
    Bus Data    Line Data    Transformer    Analysis
                                Data        Messages

  Analysis Results
```

Figure 7d. Power Flow Output Data, Sample Run Chapter 7

37

8. SYMMETRICAL SHORT CIRCUITS (CHAPTER 8)

This program computes the ac or symmetrical fault current for a bolted three-phase short circuit at any bus in an N-bus power system. The program also computes contributions to the fault current from synchronous machines, transmission lines, and transformers connected to the faulted bus. It also computes the bus voltages during the fault as well as the positive-sequence bus impedance matrix.

Input data include synchronous machine data, transmission line data, and transformer data, which are stored in three data files. The use can store the input data for up to five separate cases. The machine, line, and transformer input data files for Case 1 are called SSCM1.DAT, SSCL1.DAT, and SSCT1.DAT. Input data files for case 2 (3, 4, 5) have similar names, except the number 2 (3, 4, 5) is used.

Synchronous machine input data include the machine number, bus number, and per-unit positive-sequence reactance of each machine. When the machine reactances are subtransient reactances, the program computes subtransient fault currents. Alternatively, when the machine reactances are transient or synchronous reactances, the program computes transient or steady-state fault currents.

Transmission line data include the line number, the two buses to which the line is connected and the per-unit positive-sequence reactance for each line. Similarly, transformer input data include the transformer number, the two buses to which the transformer is connected, and the per-unit leakage reactance for each transformer.

The program user has the following three options: (1) update the input data files, (2) run the program, and (3) stop.

When the user updates the input data files, the following options for each input data file are given: (1) initialize the file (which removes all data currently stored); (2) enter or change the data for a selected machine, line or transformer; (3) remove a machine, line, or transformer; (4) display the data for one machine, line, or transformer; and (5) display the data for all machines, lines, or transformers after a given one.

When the user runs the program, the per-unit prefault voltage V_F is selected. The program then computes the positive-sequence bus impedance matrix using the one-step-at-a-time method. After Z_{bus} is computed, the fault currents, contributions to the fault currents, and bus voltages during the fault are computed for a fault at bus 1, then bus 2, . . . up to the final bus N. All computations are in per unit.

The user has the following three output display options: (1) display the fault currents, (2) display the bus voltages during the fault, and (3) display the bus-impedance matrix. The outputs can be displayed in exponential format (scientific notation) or fixed-point format (numbers with decimal points).

The following sample run computes the fault currents for the 5-bus power system whose single-line diagram is shown in Figure 7b. Machine, line, and transformer input data are given in the sample run. The prefault voltage is $V_F=1.0$ per unit.

SAMPLE RUN 8.1 CHAPTER 8 FIGURE 7b

```
ENTER YOUR CASE NUMBER ( 1,2,3,4 OR 5 ) 1

ENTER YOUR CASE NAME(ANY ALPHANUMERIC SEQUENCE)SAMPLE    RUN 8.1
SELECTIONS:

     1. UPDATE THE INPUT DATA
     2. RUN THE PROGRAM
     3. STOP


ENTER YOUR SELECTION ( 1,2 OR 3 ) 1
THESE ARE YOUR INPUT DATA SELECTIONS:

     1. UPDATE THE SYNCHRONOUS MACHINE INPUT DATA
     2. UPDATE THE TRANSMISSION LINE DATA
     3. UPDATE THE TRANSFORMER INPUT DATA
     4. DONE WITH INPUT DATA


ENTER YOUR SELECTION ( 1,2, 3 OR 4 ) 1
CHOICES :

     1. INITIALIZE FILE
     2. ADD A NEW MACHINE OR CHANGE AN EXISTING MACHINE
     3. REMOVE A MACHINE
     4. DISPLAY ONE MACHINE
     5. DISPLAY ALL MACHINES AFTER A GIVEN MACHINE
     6. DONE WITH MACHINE DATA


ENTER YOUR CHOICE ( 1,2,3,4,5 OR 6 ) : 5
ENTER THE NUMBER OF THE FIRST MACHINE TO BE DISPLAYED 1

USE Ctrl PRINT SCREEN OPTION IF YOU WANT TO PRINT THE RESULTS.

PRESS RETURN TO CONTINUE.
```

SYNCHRONOUS MACHINE INPUT DATA FOR SAMPLE RUN 8.1

MACHINE#	BUS	X1
		per unit
1	1	0.1800
2	3	0.0900

REMOVE Ctrl PRINT SCREEN AND THEN PRESS RETURN TO CONTINUE
CHOICES :

    1. INITIALIZE FILE
    2. ADD A NEW MACHINE OR CHANGE AN EXISTING MACHINE
    3. REMOVE A MACHINE
    4. DISPLAY ONE MACHINE
    5. DISPLAY ALL MACHINES AFTER A GIVEN MACHINE
    6. DONE WITH MACHINE DATA

ENTER YOUR CHOICE (1,2,3,4,5 OR 6) : 6
THESE ARE YOUR INPUT DATA SELECTIONS:

    1. UPDATE THE SYNCHRONOUS MACHINE INPUT DATA
    2. UPDATE THE TRANSMISSION LINE DATA
    3. UPDATE THE TRANSFORMER INPUT DATA
    4. DONE WITH INPUT DATA

ENTER YOUR SELECTION (1,2, 3 OR 4) 2
CHOICES :

    1. INITIALIZE FILE
    2. ADD A NEW LINE OR CHANGE AN EXISTING LINE
    3. REMOVE A LINE
    4. DISPLAY ONE LINE
    5. DISPLAY ALL LINES AFTER A GIVEN LINE
    6. DONE WITH TRANSMISSION LINE DATA

ENTER YOUR CHOICE (1,2,3,4,5 OR 6) : 5
ENTER THE NUMBER OF THE FIRST LINE TO BE DISPLAYED 1

USE THE Ctrl PRINT SCREEN OPTION TO PRINT THE RESULTS.

PRESS RETURN TO CONTINUE.

TRANSMISSION LINE INPUT DATA FOR SAMPLE RUN 8.1

LINE#	BUS	-TO- BUS	X1
			per unit
1	2	4	0.4000
2	2	4	0.4000
3	2	5	0.2000
4	2	5	0.2000
5	4	5	0.1000
6	4	5	0.1000

REMOVE Ctrl PRINT SCREEN AND THEN PRESS RETURN TO CONTINUE
CHOICES :

    1. INITIALIZE FILE
    2. ADD A NEW LINE OR CHANGE AN EXISTING LINE
    3. REMOVE A LINE
    4. DISPLAY ONE LINE
    5. DISPLAY ALL LINES AFTER A GIVEN LINE
    6. DONE WITH TRANSMISSION LINE DATA

ENTER YOUR CHOICE (1,2,3,4,5 OR 6) : 6
THESE ARE YOUR INPUT DATA SELECTIONS:

    1. UPDATE THE SYNCHRONOUS MACHINE INPUT DATA
    2. UPDATE THE TRANSMISSION LINE DATA
    3. UPDATE THE TRANSFORMER INPUT DATA
    4. DONE WITH INPUT DATA

ENTER YOUR SELECTION (1,2, 3 OR 4) 3
CHOICES :

    1. INITIALIZE FILE
    2. ADD A NEW TRANSFORMER OR CHANGE AN EXISTING TRANSFORMER
    3. REMOVE A TRANSFORMER
    4. DISPLAY ONE TRANSFORMER
    5. DISPLAY ALL TRANSFORMERS AFTER A GIVEN LINE
    6. DONE WITH TRANSFORMER DATA

ENTER YOUR CHOICE (1,2,3,4,5 OR 6) : 5
ENTER THE NUMBER OF THE FIRST TRANSFORMER TO BE DISPLAYED 1

USE THE Ctrl PRINT SCREEN OPTION IF YOU WANT TO PRINT THE RESULTS.

PRESS RETURN TO CONTINUE.

        TRANSFORMER INPUT DATA FOR SAMPLE RUN 8.1

TRANSF#	BUS-TO-BUS		XL
			per unit
1	1	5	0.0800
2	3	4	0.0400

REMOVE Ctrl PRINT SCREEN AND THEN PRESS RETURN TO CONTINUE

```
CHOICES :

        1. INITIALIZE FILE
        2. ADD A NEW TRANSFORMER OR CHANGE AN EXISTING TRANSFORMER
        3. REMOVE A TRANSFORMER
        4. DISPLAY ONE TRANSFORMER
        5. DISPLAY ALL TRANSFORMERS AFTER A GIVEN LINE
        6. DONE WITH TRANSFORMER DATA


ENTER YOUR CHOICE ( 1,2,3,4,5 OR 6 ) : 6
THESE ARE YOUR INPUT DATA SELECTIONS:

        1. UPDATE THE SYNCHRONOUS MACHINE INPUT DATA
        2. UPDATE THE TRANSMISSION LINE DATA
        3. UPDATE THE TRANSFORMER INPUT DATA
        4. DONE WITH INPUT DATA


ENTER YOUR SELECTION ( 1,2, 3 OR 4 ) 4
SELECTIONS:

        1. UPDATE THE INPUT DATA
        2. RUN THE PROGRAM
        3. STOP


ENTER YOUR SELECTION ( 1,2 OR 3 ) 2




WHAT IS THE PREFAULT VOLTAGE ( per unit )? 1.0

DISPLAY THE FAULT CURRENTS (Y OR N)? Y

DISPLAY THE BUS VOLTAGES DURING THE FAULT (Y OR N)? Y

DISPLAY THE BUS IMPEDANCE MATRICES (Y OR N)? Y

DO YOU WANT THE OUTPUTS PRINTED IN EXPONENTIAL FORMAT ( Y OR N )  ? N

DO YOU WANT TO CONTINUE (Y) OR RESET THE DATA (N) ? Y

USE THE  Ctrl PRINT SCREEN OPTION NOW IF YOU WANT TO PRINT THE RESULTS.
PRESS RETURN TO CONTINUE
```

FAULT CURRENTS FOR SAMPLE RUN 8.1

FAULT BUS	THREE-PHASE FAULT CURRENT per unit	GEN LINE OR TRSF	BUS-TO-BUS			CONTRIBUTIONS TO FAULT CURRENT per unit
1	9.510					
		G 1	GRND	–	1	5.5556
		T 1	5	–	1	3.9548
2	6.344					
		L 1	4	–	2	1.1935
		L 2	4	–	2	1.1935
		L 3	5	–	2	1.9786
		L 4	5	–	2	1.9786
3	14.028					
		G 2	GRND	–	3	11.1111
		T 2	4	–	3	2.9167
4	10.994					
		L 1	2	–	4	0.2358
		L 2	2	–	4	0.2358
		L 5	5	–	4	1.4151
		L 6	5	–	4	1.4151
		T 2	3	–	4	7.6923
5	9.631					
		L 3	2	–	5	0.4132
		L 4	2	–	5	0.4132
		L 5	4	–	5	2.4793
		L 6	4	–	5	2.4793
		T 1	1	–	5	3.8462

```
VF =    1            PER-UNIT BUS VOLTAGES DURING FAULTS

                              SAMPLE RUN 8.1
FAULT
 BUS            BUS1(6..)  BUS2(7..)  BUS3(8..)  BUS4(9..)  BUS5(10..)


   1             0.0000     0.3729     0.6441     0.4859     0.3164.
   2             0.5817     0.0000     0.6382     0.4774     0.3957
   3             0.4750     0.2000     0.0000     0.1167     0.2417
   4             0.4057     0.0943     0.3077     0.0000     0.1415
   5             0.3077     0.0826     0.4793     0.2479     0.0000




                   PER-UNIT ZBUS FOR SAMPLE RUN 8.1


 ROW 1           0.10515    0.06594    0.03743    0.05406    0.07188
 ROW 2           0.06594    0.15762    0.05703    0.08238    0.09525
 ROW 3           0.03743    0.05703    0.07129    0.06297    0.05406
 ROW 4           0.05406    0.08238    0.06297    0.09096    0.07809
 ROW 5           0.07188    0.09525    0.05406    0.07809    0.10383

REMOVE THE Ctrl PRINT SCREEN OPTION AND PRESS RETURN TO CONTINUE




    SELECTIONS:
            1. UPDATE THE INPUT DATA
            2. RUN THE PROGRAM
            3. STOP


    ENTER YOUR SELECTION ( 1,2 OR 3 ) 3


    YOU CAN SAVE THE INPUT DATA FOR UP TO FIVE SEPARATE CASES.

    ENTER YOUR CASE NUMBER ( 1,2,3,4 OR 5 ) 1

    ENTER YOUR CASE NAME(ANY ALPHANUMERIC SEQUENCE)
    SELECTIONS:
            1. UPDATE THE INPUT DATA
            2. RUN THE PROGRAM
            3. STOP


    ENTER YOUR SELECTION ( 1,2 OR 3 ) 3
```

The following sample run computes the bus impedance matrix for the 3-bus circuit shown in Figure 8a. The Z_{bus} matrix displayed in the output is the same as that given in Problem 8.11 of the text *Power System Analysis and Design with Personal Computer Applications*. After computing Z_{bus} from this sample run, the user can then work Problem 8.22 in the text.

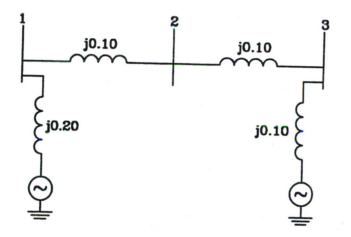

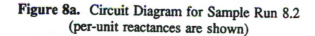

Figure 8a. Circuit Diagram for Sample Run 8.2
(per-unit reactances are shown)

45

SAMPLE RUN 8.2 CHAPTER 8 FIGURE 8a

ENTER YOUR CASE NAME(ANY ALPHANUMERIC SEQUENCE)SAMPLE RUN 8.2
SELECTIONS:

    1. UPDATE THE INPUT DATA
    2. RUN THE PROGRAM
    3. STOP

ENTER YOUR SELECTION (1,2 OR 3) 1
THESE ARE YOUR INPUT DATA SELECTIONS:

    1. UPDATE THE SYNCHRONOUS MACHINE INPUT DATA
    2. UPDATE THE TRANSMISSION LINE DATA
    3. UPDATE THE TRANSFORMER INPUT DATA
    4. DONE WITH INPUT DATA

ENTER YOUR SELECTION (1,2, 3 OR 4) 1
CHOICES :

    1. INITIALIZE FILE
    2. ADD A NEW MACHINE OR CHANGE AN EXISTING MACHINE
    3. REMOVE A MACHINE
    4. DISPLAY ONE MACHINE
    5. DISPLAY ALL MACHINES AFTER A GIVEN MACHINE
    6. DONE WITH MACHINE DATA

ENTER YOUR CHOICE (1,2,3,4,5 OR 6) : 5
ENTER THE NUMBER OF THE FIRST MACHINE TO BE DISPLAYED 1

USE Ctrl PRINT SCREEN OPTION IF YOU WANT TO PRINT THE RESULTS.

PRESS RETURN TO CONTINUE.

SYNCHRONOUS MACHINE INPUT DATA FOR SAMPLE RUN 8.2

MACHINE#	BUS	X1
		per unit
1	1	0.2000
2	3	0.1000

REMOVE Ctrl PRINT SCREEN AND THEN PRESS RETURN TO CONTINUE

CHOICES :

    1. INITIALIZE FILE
    2. ADD A NEW MACHINE OR CHANGE AN EXISTING MACHINE
    3. REMOVE A MACHINE
    4. DISPLAY ONE MACHINE
    5. DISPLAY ALL MACHINES AFTER A GIVEN MACHINE
    6. DONE WITH MACHINE DATA

ENTER YOUR CHOICE (1,2,3,4,5 OR 6) : 6
THESE ARE YOUR INPUT DATA SELECTIONS:

    1. UPDATE THE SYNCHRONOUS MACHINE INPUT DATA
    2. UPDATE THE TRANSMISSION LINE DATA
    3. UPDATE THE TRANSFORMER INPUT DATA
    4. DONE WITH INPUT DATA

ENTER YOUR SELECTION (1,2, 3 OR 4) 2
CHOICES :

    1. INITIALIZE FILE
    2. ADD A NEW LINE OR CHANGE AN EXISTING LINE
    3. REMOVE A LINE
    4. DISPLAY ONE LINE
    5. DISPLAY ALL LINES AFTER A GIVEN LINE
    6. DONE WITH TRANSMISSION LINE DATA

ENTER YOUR CHOICE (1,2,3,4,5 OR 6) : 5
ENTER THE NUMBER OF THE FIRST LINE TO BE DISPLAYED 1

USE THE Ctrl PRINT SCREEN OPTION TO PRINT THE RESULTS.

PRESS RETURN TO CONTINUE.

TRANSMISSION LINE INPUT DATA FOR SAMPLE RUN 8.2

LINE#	BUS	-TO- BUS	X1
			per unit
1	1	2	0.1000
2	2	3	0.1000

REMOVE Ctrl PRINT SCREEN AND THEN PRESS RETURN TO CONTINUE

CHOICES :

    1. INITIALIZE FILE
    2. ADD A NEW LINE OR CHANGE AN EXISTING LINE
    3. REMOVE A LINE
    4. DISPLAY ONE LINE
    5. DISPLAY ALL LINES AFTER A GIVEN LINE
    6. DONE WITH TRANSMISSION LINE DATA

ENTER YOUR CHOICE (1,2,3,4,5 OR 6) : 6

THESE ARE YOUR INPUT DATA SELECTIONS:

    1. UPDATE THE SYNCHRONOUS MACHINE INPUT DATA
    2. UPDATE THE TRANSMISSION LINE DATA
    3. UPDATE THE TRANSFORMER INPUT DATA
    4. DONE WITH INPUT DATA

ENTER YOUR SELECTION (1,2, 3 OR 4) 4

SELECTIONS:

    1. UPDATE THE INPUT DATA
    2. RUN THE PROGRAM
    3. STOP

ENTER YOUR SELECTION (1,2 OR 3) 2

WHAT IS THE PREFAULT VOLTAGE (per unit)? 1.0

DISPLAY THE FAULT CURRENTS (Y OR N)? N

DISPLAY THE BUS VOLTAGES DURING THE FAULT (Y OR N)? N

DISPLAY THE BUS IMPEDANCE MATRICES (Y OR N)? Y

DO YOU WANT THE OUTPUTS PRINTED IN EXPONENTIAL FORMAT (Y OR N) ? N

DO YOU WANT TO CONTINUE (Y) OR RESET THE DATA (N) ? Y

USE THE Ctrl PRINT SCREEN OPTION NOW IF YOU WANT TO PRINT THE RESULTS.
PRESS RETURN TO CONTINUE

                        PER-UNIT ZBUS FOR SAMPLE RUN 8.2

        ROW 1   0.12000   0.08000   0.04000
        ROW 2   0.08000   0.12000   0.06000
        ROW 3   0.04000   0.06000   0.08000

REMOVE THE Ctrl PRINT SCREEN OPTION AND PRESS RETURN TO CONTINUE
SELECTIONS:

    1. UPDATE THE INPUT DATA
    2. RUN THE PROGRAM
    3. STOP

ENTER YOUR SELECTION (1,2 OR 3) 3

9. SHORT CIRCUITS (CHAPTER 9)

This program computes the ac or symmetrical fault current for one of the following faults in an N-bus power system: balanced three-phase fault, single line-to-ground fault through fault impedance Z_F, line-to-line fault through Z_F, or double line-to-ground fault through Z_F. The program also computes the contributions to the fault current from synchronous machines, transmission lines, and transformers connected to the faulted bus. And it computes the three-phase bus voltages during the fault as well as the zero-, positive-, and negative-sequence bus impedance matrices.

Input data include synchronous machine data, transmission line data, and transformer data, which are stored in three data files. The user can store the input data for up to five separate cases. The machine, line, and transformer input data files for case 1 are called SCM1.DAT, SCL1.DAT, and SCT1.DAT. Input data fields for Case 2 (3, 4, 5) have similar names, except the number 2 (3, 4, 5) is used.

Synchronous machine input data include machine number, bus number, per-unit zero-, positive-, and negative-sequence machine reactances, and per-unit neutral reactance for each machine. When the positive-sequence machine reactances are subtransient reactances, the program computes subtransient fault currents. Alternatively, when the positive-sequence machine reactances are transient or synchronous reactances, the program computes transient or steady-state fault currents.

Transmission line input data include line number, the two buses to which the line is connected, and the per-unit zero- and positive-sequence reactance for each line. Transformer input data include transformer number, low-voltage bus number, low-voltage connection (delta or wye), high-voltage bus number, high-voltage connection, per-unit leakage reactance, and per-unit neutral reactance of each wye-connection for each transformer.

For a solidly grounded neutral, enter 0 (zero) for neutral reactance in the machine or transformer input data. For an approximation to an open neutral, enter a large value such as 1000 for neutral reactance.

The program user has the following three selections: (1) update the input data files, (2) run the program, and (3) stop.

When the user updates the input data files, the following options for each input data file are given: (1) initialize the file (which removes all data currently stored); (2) enter or change the data for a selected machine, line, or transformer; (3) remove a machine, line, or transformer; (4) display the data for one machine, line, or transformer; and (5) display the data for all machines, lines, or transformers after a given one.

When the user runs the program, one of the following four faults is selected: (1) three-phase fault, (2) single line-to-ground fault through Z_F, (3) line-to-line fault through Z_F, or

(4) double line-to-ground fault through Z_F. The user also selects the per-unit resistive and reactive components of the fault impedance, and the per-unit prefault voltage V_F.

The program then computes the zero-, positive-, and negative-sequence bus impedance matrices using the one step-at-a-time method. After Z_{bus0}, Z_{bus1}, and Z_{bus2} are computed, the fault currents, contributions to the fault currents, and bus voltages during the fault are computed for a fault at bus 1, then bus 2, ... up to the final bus N. All computations are in per unit.

The user has three output options: (1) display the fault currents and contributions to the fault currents; (2) display the three-phase bus voltages during the faults; and (3) display the zero-, positive-, and negative-sequence bus impedance matrices. The outputs can be displayed in exponential format (scientific notation) or fixed-point format (numbers with decimal points).

The following sample run computes the fault currents for the 5-bus power system whose single-line diagram is shown in Figure 7b. Machine, line, and transformer input data are given in the sample run. A double line-to-ground fault with $Z_F=0$ and $V_F=1.0$ is selected.

SAMPLE RUN CHAPTER 9 FIGURE 7b

```
ENTER YOUR CASE NUMBER ( 1,2,3,4 OR 5 ) 1

ENTER YOUR CASE NAME(ANY ALPHANUMERIC SEQUENCE)SAMPLE RUN
SELECTIONS:

     1. UPDATE THE INPUT DATA
     2. RUN THE PROGRAM
     3. STOP


ENTER YOUR SELECTION ( 1,2 OR 3 ) 1
THESE ARE YOUR INPUT DATA SELECTIONS:

     1. UPDATE THE SYNCHRONOUS MACHINE INPUT DATA
     2. UPDATE THE TRANSMISSION LINE DATA
     3. UPDATE THE TRANSFORMER INPUT DATA
     4. DONE WITH INPUT DATA


ENTER YOUR SELECTION ( 1,2, 3 OR 4 ) 1
CHOICES :

     1. INITIALIZE FILE
     2. ADD A NEW MACHINE OR CHANGE AN EXISTING MACHINE
     3. REMOVE A MACHINE
     4. DISPLAY ONE MACHINE
     5. DISPLAY ALL MACHINES AFTER A GIVEN MACHINE
     6. DONE WITH MACHINE DATA


ENTER YOUR CHOICE ( 1,2,3,4,5 OR 6 ) : 5
ENTER THE NUMBER OF THE FIRST MACHINE TO BE DISPLAYED 1
```

USE Ctrl PRINT SCREEN OPTION IF YOU WANT TO PRINT THE RESULTS.

PRESS RETURN TO CONTINUE.

SYNCHRONOUS MACHINE INPUT DATA FOR SAMPLE RUN

MACHINE#	BUS	X0 per unit	X1 per unit	X2 per unit	XN per unit
1	1	0.0500	0.1800	0.1800	0.0000
2	3	0.0200	0.0900	0.0900	0.0100

REMOVE Ctrl PRINT SCREEN AND THEN PRESS RETURN TO CONTINUE
CHOICES :

   1. INITIALIZE FILE
   2. ADD A NEW MACHINE OR CHANGE AN EXISTING MACHINE
   3. REMOVE A MACHINE
   4. DISPLAY ONE MACHINE
   5. DISPLAY ALL MACHINES AFTER A GIVEN MACHINE
   6. DONE WITH MACHINE DATA

ENTER YOUR CHOICE (1,2,3,4,5 OR 6) : 6

THESE ARE YOUR INPUT DATA SELECTIONS:

   1. UPDATE THE SYNCHRONOUS MACHINE INPUT DATA
   2. UPDATE THE TRANSMISSION LINE DATA
   3. UPDATE THE TRANSFORMER INPUT DATA
   4. DONE WITH INPUT DATA

ENTER YOUR SELECTION (1,2, 3 OR 4) 2
CHOICES :

   1. INITIALIZE FILE
   2. ADD A NEW LINE OR CHANGE AN EXISTING LINE
   3. REMOVE A LINE
   4. DISPLAY ONE LINE
   5. DISPLAY ALL LINES AFTER A GIVEN LINE
   6. DONE WITH TRANSMISSION LINE DATA

ENTER YOUR CHOICE (1,2,3,4,5 OR 6) : 5
ENTER THE NUMBER OF THE FIRST LINE TO BE DISPLAYED 1

USE THE Ctrl PRINT SCREEN OPTION TO PRINT THE RESULTS.

PRESS RETURN TO CONTINUE.

TRANSMISSION LINE INPUT DATA FOR SAMPLE RUN

LINE#	BUS -TO- BUS		X0 per unit	X1 per unit
1	2	4	1.2000	0.4000
2	2	4	1.2000	0.4000
3	2	5	0.6000	0.2000
4	2	5	0.6000	0.2000
5	4	5	0.3000	0.1000
6	4	5	0.3000	0.1000

REMOVE Ctrl PRINT SCREEN AND THEN PRESS RETURN TO CONTINUE
CHOICES :

    1. INITIALIZE FILE
    2. ADD A NEW LINE OR CHANGE AN EXISTING LINE
    3. REMOVE A LINE
    4. DISPLAY ONE LINE
    5. DISPLAY ALL LINES AFTER A GIVEN LINE
    6. DONE WITH TRANSMISSION LINE DATA

ENTER YOUR CHOICE (1,2,3,4,5 OR 6) : 6
THESE ARE YOUR INPUT DATA SELECTIONS:

    1. UPDATE THE SYNCHRONOUS MACHINE INPUT DATA
    2. UPDATE THE TRANSMISSION LINE DATA
    3. UPDATE THE TRANSFORMER INPUT DATA
    4. DONE WITH INPUT DATA

ENTER YOUR SELECTION (1,2, 3 OR 4) 3

CHOICES :

    1. INITIALIZE FILE
    2. ADD A NEW TRANSFORMER OR CHANGE AN EXISTING TRANSFORMER
    3. REMOVE A TRANSFORMER
    4. DISPLAY ONE TRANSFORMER
    5. DISPLAY ALL TRANSFORMERS AFTER A GIVEN ONE
    6. DONE WITH TRANSFORMER DATA

ENTER YOUR CHOICE (1,2,3,4,5 OR 6) : 5
ENTER THE NUMBER OF THE FIRST TRANSFORMER TO BE DISPLAYED 1

USE THE Ctrl PRINT SCREEN OPTION IF YOU WANT TO PRINT THE RESULTS.

PRESS RETURN TO CONTINUE.

TRANSFORMER INPUT DATA FOR SAMPLE RUN

TRANSF#	LOW (CNC) VOLT. BUS		HIGH(CNC) VOLT. BUS		XL	NEUTRAL REACTANCES	
						LV	HV
					per unit	per unit	per unit
1	1	(D)	5	(Y)	0.0800		0.0000
2	3	(D)	4	(Y)	0.0400		0.0000

REMOVE Ctrl PRINT SCREEN AND THEN PRESS RETURN TO CONTINUE
CHOICES :

    1. INITIALIZE FILE
    2. ADD A NEW TRANSFORMER OR CHANGE AN EXISTING TRANSFORMER
    3. REMOVE A TRANSFORMER
    4. DISPLAY ONE TRANSFORMER
    5. DISPLAY ALL TRANSFORMERS AFTER A GIVEN ONE
    6. DONE WITH TRANSFORMER DATA

ENTER YOUR CHOICE (1,2,3,4,5 OR 6) : 6
THESE ARE YOUR INPUT DATA SELECTIONS:

    1. UPDATE THE SYNCHRONOUS MACHINE INPUT DATA
    2. UPDATE THE TRANSMISSION LINE DATA
    3. UPDATE THE TRANSFORMER INPUT DATA
    4. DONE WITH INPUT DATA

ENTER YOUR SELECTION (1,2, 3 OR 4) 4
SELECTIONS:

    1. UPDATE THE INPUT DATA
    2. RUN THE PROGRAM
    3. STOP

ENTER YOUR SELECTION (1,2 OR 3) 2

ONE OF THE FOLLOWING FAULT TYPES IS TO BE SELECTED:
    (1) THREE-PHASE FAULT
    (2) SINGLE LINE-TO-GROUND FAULT THROUGH ZF
    (3) LINE-TO-LINE FAULT THROUGH ZF
    (4) DOUBLE LINE-TO-GROUND FAULT THROUGH ZF

ENTER THE FAULT TYPE (1,2,3 OR 4) 4

WHAT IS THE PREFAULT VOLTAGE (per unit) ? 1.0

WHAT IS THE RESISTIVE PART OF THE FAULT IMPEDANCE(per unit)? 0

WHAT IS THE REACTIVE PART OF THE FAULT IMPEDANCE(per unit)? 0

DISPLAY THE FAULT CURRENTS (Y OR N)? Y

DISPLAY THE BUS VOLTAGES DURING THE FAULT (Y OR N)? Y

DISPLAY THE BUS IMPEDANCE MATRICES (Y OR N)? Y

DO YOU WANT THE OUTPUTS PRINTED IN EXPONENTIAL FORMAT (Y OR N) ? N

DO YOU WANT TO CONTINUE (Y) OR RESET THE DATA (N) ? Y

USE THE Ctrl PRINT SCREEN OPTION NOW IF YOU WANT TO PRINT THE RESULTS.
PRESS RETURN TO CONTINUE

VF= 1 /0   BUS VOLTAGES DURING FAULTS FOR SAMPLE RUN

FAULT BUS	BUS #	PHASE A	PHASE B per unit/degrees	PHASE C
1	1	0.7312/ 0.00	0.0000/ 0.00	0.0000/ 0.00
	2	0.6165/ 17.60	0.3729/-90.00	0.6165/162.40
	3	0.8176/ 0.00	0.6915/233.76	0.6915/126.24
	4	0.6825/ 20.85	0.4859/270.00	0.6825/159.15
	5	0.5844/ 15.71	0.3164/-90.00	0.5844/164.29
2	1	0.8274/-20.58	0.8274/200.58	0.5817/ 90.00
	2	1.1216/ 0.00	0.0000/ 8.21	0.0000/ -8.21
	3	0.8492/-22.07	0.8492/202.07	0.6382/ 90.00
	4	0.8998/ 0.00	0.5770/225.77	0.5770/134.23
	5	0.9123/ 0.00	0.4963/223.67	0.4963/136.33
3	1	0.7815/ 0.00	0.5674/226.47	0.5674/133.53
	2	0.5863/ 9.82	0.2000/-90.00	0.5863/170.18
	3	0.8757/ 0.00	0.0000/ 0.00	0.0000/ 0.00
	4	0.5507/ 6.08	0.1167/-90.00	0.5507/173.92
	5	0.6049/ 11.52	0.2417/-90.00	0.6049/168.48
4	1	0.6049/-19.59	0.6049/199.59	0.4057/ 90.00
	2	0.6040/ 0.00	0.1406/215.53	0.1406/144.47
	3	0.5433/-16.45	0.5433/196.45	0.3077/ 90.00
	4	0.6369/ 0.00	0.0000/180.00	0.0000/180.00
	5	0.5875/ 0.00	0.2109/215.53	0.2109/144.47
5	1	0.5931/-15.03	0.5931/195.03	0.3077/ 90.00
	2	0.7420/ 0.00	0.1112/220.04	0.1112/139.96
	3	0.6886/-20.37	0.6886/200.37	0.4793/ 90.00
	4	0.6929/ 0.00	0.3337/220.04	0.3337/139.96
	5	0.7665/ 0.00	0.0000/ 0.00	0.0000/ 0.00

FAULT CURRENTS FOR SAMPLE RUN

VF = 1 /0   ZF = 0 + j 0

FAULT BUS	DOUBLE LINE-TO-GROUND FAULT CURRENT (PHASE B) per unit/degrees	GEN LINE OR TRSF	BUS-TO-BUS	CONTRIBUTIONS TO FAULT CURRENT		
				PHASE A	PHASE B	PHASE C
				per unit/degrees		
1	11.014/138.40					
		G 1	GRND - 1	2.0270/ 90.00	7.9257/ 127.38	7.9257/ 52.62
		T 1	5 - 1	2.0270/ -90.00	3.5718/ 163.52	3.5718/ 16.48
2	5.996/156.39					
		L 1	4 - 2	0.0159/ -90.00	1.1217/ 157.14	1.1217/ 22.86
		L 2	4 - 2	0.0159/ -90.00	1.1217/ 157.14	1.1217/ 22.86
		L 3	5 - 2	0.0159/ 90.00	1.8765/ 155.95	1.8765/ 24.05
		L 4	5 - 2	0.0159/ 90.00	1.8765/ 155.95	1.8765/ 24.05
3	14.976/144.21					
		G 2	GRND - 3	1.2139/ 90.00	12.6103/ 139.74	12.6103/ 40.26
		T 2	4 - 3	1.2139/ -90.00	2.5978/ 166.49	2.5978/ 13.51
4	13.442/135.10					
		L 1	2 - 4	0.0630/ -90.00	0.2479/ 145.47	0.2479/ 34.53
		L 2	2 - 4	0.0630/ -90.00	0.2479/ 145.47	0.2479/ 34.53
		L 5	5 - 4	0.3779/ -90.00	1.4877/ 145.47	1.4877/ 34.53
		L 6	5 - 4	0.3779/ -90.00	1.4877/ 145.47	1.4877/ 34.53
		T 2	3 - 4	0.8819/ 90.00	10.0470/ 131.53	10.0470/ 48.47
5	10.931/139.74					
		L 3	2 - 5	0.0938/ -90.00	0.4146/ 149.68	0.4146/ 30.32
		L 4	2 - 5	0.0938/ -90.00	0.4146/ 149.68	0.4146/ 30.32
		L 5	4 - 5	0.5628/ -90.00	2.4874/ 149.68	2.4874/ 30.32
		L 6	4 - 5	0.5628/ -90.00	2.4874/ 149.68	2.4874/ 30.32
		T 1	1 - 5	1.3131/ 90.00	5.3091/ 128.86	5.3091/ 51.14

Zbus MATRICES FOR SAMPLE RUN

ZF = 0 + j 0   per unit

Zbus 0

ROW 1	0.05000	0.00000	0.00000	0.00000	0.00000
ROW 2	0.00000	0.23356	0.00000	0.01977	0.04046
ROW 3	0.00000	0.00000	0.05000	0.00000	0.00000
ROW 4	0.00000	0.01977	0.00000	0.03356	0.01287
ROW 5	0.00000	0.04046	0.00000	0.01287	0.05425

Zbus 1

ROW 1	0.10515	0.06594	0.03743	0.05406	0.07188
ROW 2	0.06594	0.15762	0.05703	0.08238	0.09525
ROW 3	0.03743	0.05703	0.07129	0.06297	0.05406
ROW 4	0.05406	0.08238	0.06297	0.09096	0.07809
ROW 5	0.07188	0.09525	0.05406	0.07809	0.10383

Zbus 2

ROW 1	0.10515	0.06594	0.03743	0.05406	0.07188
ROW 2	0.06594	0.15762	0.05703	0.08238	0.09525
ROW 3	0.03743	0.05703	0.07129	0.06297	0.05406
ROW 4	0.05406	0.08238	0.06297	0.09096	0.07809
ROW 5	0.07188	0.09525	0.05406	0.07809	0.10383

REMOVE THE Ctrl PRINT SCREEN OPTION AND PRESS RETURN TO CONTINUE
SELECTIONS:

    1. UPDATE THE INPUT DATA
    2. RUN THE PROGRAM
    3. STOP

ENTER YOUR SELECTION (1,2 OR 3) 3

10.   INTENTIONALLY LEFT BLANK

11.   INTENTIONALLY LEFT BLANK

(There is no software for these Chapters.)

12. TRANSMISSION LINE TRANSIENTS (CHAPTER 12)

This program computes voltage transients for the single-phase circuit shown in Figure 12a. This circuit consists of 3 single-phase lossless line sections, 10 buses, 4 independent current sources, and 10 lumped parallel RLC elements. The program computes the 10 bus voltages at discrete times Δt, $2\Delta t$, . . . , $N\Delta t$.

Input data include line data, RLC data, and independent current source data, which are stored in three separate data files. The user can store up to five separate cases. The line, RLC, and source data files for Case 1 are called TLT1.DAT, TLTRLC1.DAT, and TLTS1.DAT. Input data files for Case 2 (3, 4, 5) have similar names, except the number 2 (3, 4, 5) is used.

The user has the following three options for setting circuit breakers B1 and B2 in Figure 12a:
(1) B1 and B2 open - The first 4 buses are considered along with line section 1, the first 4 RLC elements, and the first 2 sources. The remainder of the circuit is disregarded.

(2) B1 closed and B2 open - The first 7 buses are considered along with line sections 1 and 2, the first 7 RLC elements, and the first 3 sources. The remainder of the circuit is disregarded.

(3) B1 and B2 closed - The entire 10-bus circuit is considered.

Line input data include the characteristic impedance $Z_c(\Omega)$ and transit time tau (μs) for each lossless line section.

A lossy line section can be modeled by lumping the series line resistance and including half of this lumped resistance in the R element on each side of the line section.

RLC input data include the values of 1/R (1/ohms), 1/L (1/henries) and C (farads) for each parallel RLC element. Two values along with a "changeover voltage" (kV) are entered for each element As shown in Figure 12.b, the program uses the first value when the voltage across the element is less than the changeover voltage, and the second value when the voltage is greater than or equal to the changeover voltage. In this way, voltage-dependent, piecewise constant RLC elements can be modelled.

The user can select an open circuit for any RLC element by entering zero values for 1/R, 1/L, and C. Also, the use can approximate a short circuit for any RLC element by entering a large value for 1/R (for example 1/R = 10) and zero values for 1/L and C. However, do not select too high a value for 1/R, otherwise the computation method may break down.

The user has the following four options for each independent current source, as shown in Figure 12b: (S) square wave; (R) ramp; (T) triangular wave; or (E) double exponential wave. Source input data include the source type (S, R, T, or E), source magnitude I (kA or kA/s); and the time T1 and T2 (μs) for each independent current source.

The user has the following three options: (1) update the input data files; (2) run the program; or (3) stop.

When the user updates the input data files, the following options for each input data file are given: (1) create a new file; (2) change the date for one line section, RLC element, or source; or (3) display the file.

When the user runs the program, the calculation time interval Δt (μs) and the final time t_F(μs) are selected. Δt should be small in order to obtain an accurate solution. One guideline is to select Δt to be 1/10 of either the smallest transit time tau or the smallest time constant of your circuit. However, a program restriction is that the ratio tau/Δt must be less than 250 for each line section. That is, the program will not allow too small a value of Δt.

The program replaces the circuit elements in Figure 12a by their discrete-time equivalent circuits. Then nodal analysis is used to compute the 10 bus voltages at discrete time intervals Δt, 2Δt, . . . , until NΔt=t_F. Gauss elimination is used to solve the nodal equations.

The user selects up to 5 bus voltages for display during any run. The user also selects the printout integer K. Then the bus voltages at time intervals of KΔt are displayed. Also, the user can display the bus voltages in exponential format (scientific notation) or fixed-point format (numbers with decimal points).

The following sample run solves Example 12.10 of the text *Power System Analysis and Design with Personal Computer Applications*.

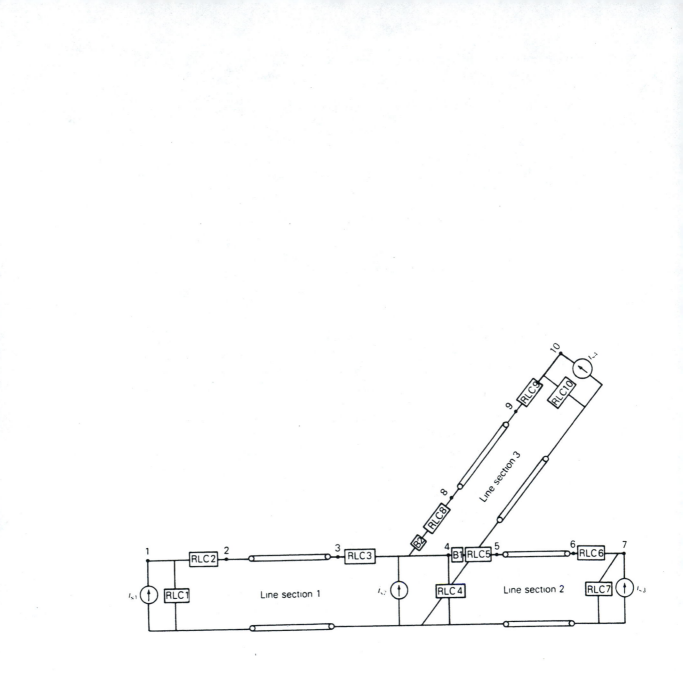

Figure 12a. Circuit Diagram for Chapter 12
Transmission Line Transients

60

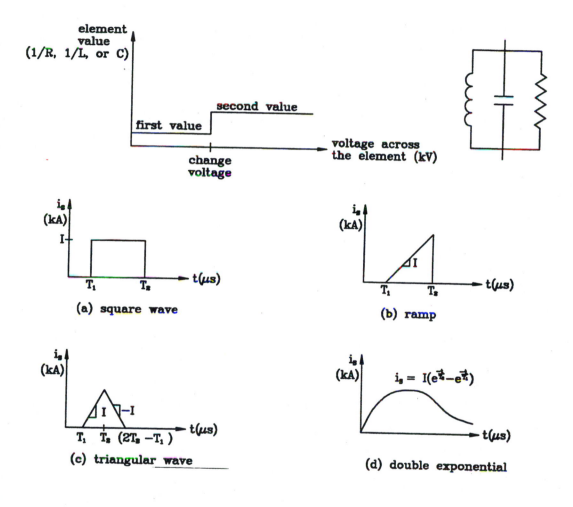

element
value
(1/R, 1/L, or C)

second value

first value

change
voltage

voltage across
the element (kV)

i_s
(kA)

I

T_1 T_2 $t(\mu s)$

(a) square wave

i_s
(kA)

T_1 T_2 $t(\mu s)$

$\triangle I$

(b) ramp

i_s
(kA)

$\triangle$ I $\triangledown$ −I

T_1 T_2 $(2T_2 - T_1)$ $t(\mu s)$

(c) triangular wave

i_s
(kA)

$i_s = I(e^{\frac{-t}{a}} - e^{\frac{-t}{b}})$

$t(\mu s)$

(d) double exponential

Figure 12b. RLC element values and independent current sources for Chapter 12

CHAPTER 12 SAMPLE RUN EXAMPLE 12.10

WHAT IS YOUR CASE NUMBER (1,2,3,4 OR 5) ? 1

ENTER YOUR CASE NAME EXAMPLE 12.10

IS CIRCUIT BREAKER B1 CLOSED (Y OR N) ? Y

IS CIRCUIT BREAKER B2 CLOSED (Y OR N) ? N

WOULD YOU LIKE TO:
        1. CREATE, EDIT OR LOOK AT THE INPUT DATA FILES
        2. RUN THE PROGRAM
        3. STOP

ENTER YOUR SELECTION(1,2, OR 3) 2
ENTER THE CALCULATION TIME INTERVAL (MICROSECONDS) 0.1

ENTER THE FINAL TIME (MICROSECONDS) 150

YOU MAY DISPLAY FROM 1 TO 5 BUS VOLTAGE OUTPUTS.

ENTER THE NUMBER OF OUTPUTS 3

ENTER OUTPUT 1 BUS NUMBER
? 1

ENTER OUTPUT 2 BUS NUMBER
? 4

ENTER OUTPUT 3 BUS NUMBER
? 7
THE OUTPUTS ARE PRINTED EVERY KTH TIME INTERVAL

ENTER THE PRINTOUT INTEGER K(ANY INTEGER>=1)30

DO YOU WANT THE OUTPUTS PRINTED IN EXPONENTIAL FORMAT (Y OR N)? N

DO YOU WANT TO PRINT THE INPUT DATA (Y OR N) ? Y

DO YOU WANT TO CONTINUE (Y) OR RESET THE DATA (N) ? Y

USE THE Ctrl PRINT SCREEN NOW IF YOU WANT A PRINTED OUTPUT

PRESS RETURN TO CONTINUE

INPUT DATA FOR EXAMPLE 12.10

LINE DATA FOR EXAMPLE 12.10

LINE SECTION	CHARACTERISTIC IMPEDANCE	TRANSIT TIME
	ohms	micro seconds
1	300	16.6667
2	300	16.6667

CIRCUIT BREAKER ONE IS CLOSED

CIRCUIT BREAKER TWO IS OPEN

RLC DATA FOR EXAMPLE 12.10

RLC ELEMENT	1/R				1/L			C		
	FIRST VALUE	SECOND VALUE	CHANGE VOLTAGE	FIRST VALUE	SECOND VALUE	CHANGE VOLTAGE	FIRST VALUE	SECOND VALUE	CHANGE VOLTAGE
	1/ohms	1/ohms	kV	1/henry	1/henry	kV	farads	farads	kV
1	0.0E+00	0.0E+00	0.0E+00	0.0E+00	0.0E+00	0.0E+00	0.0E+00	0.0E+00	0.0E+00
2	8.0E+00	8.0E+00	0.0E+00	0.0E+00	0.0E+00	0.0E+00	0.0E+00	0.0E+00	0.0E+00
3	8.0E+00	8.0E+00	0.0E+00	0.0E+00	0.0E+00	0.0E+00	0.0E+00	0.0E+00	0.0E+00
4	0.0E+00	0.0E+00	0.0E+00	0.0E+00	0.0E+00	0.0E+00	0.0E+00	0.0E+00	0.0E+00
5	8.0E+00	8.0E+00	0.0E+00	0.0E+00	0.0E+00	0.0E+00	0.0E+00	0.0E+00	0.0E+00
6	8.0E+00	8.0E+00	0.0E+00	0.0E+00	0.0E+00	0.0E+00	0.0E+00	0.0E+00	0.0E+00
7	5.0E-07	2.2E-01	5.5E+01	0.0E+00	0.0E+00	0.0E+00	6.0E-09	6.0E-09	0.0E+00

INDEPENDENT CURRENT SOURCE DATA FOR EXAMPLE 12.10

SOURCE	TYPE	I	T1	T2
		kA or kA/S	micro seconds	micro seconds
1	S	0	0	0
2	S	20	0	20
3	S	0	0	0

OUTPUT DATA FOR EXAMPLE 12.10

BUS VOLTAGES

TIME micro seconds	V 1 kV	V 4 kV	V 7 kV
0.000	0.00	3001.29	0.00
3.000	0.00	3001.29	0.00
6.000	0.00	3001.29	0.00
9.000	0.00	3001.29	0.00
12.000	0.00	3001.29	0.00
15.000	0.00	3001.29	0.00
18.000	6000.52	3001.29	89.52
21.000	6000.52	0.00	89.52
24.000	6000.52	0.00	89.52
27.000	6000.52	0.00	89.52
30.000	6000.52	0.00	89.52
33.000	6000.52	0.00	89.52
36.000	6000.52	92.41	89.52
39.000	0.00	92.41	-7.30
42.000	0.00	92.41	-1.38
45.000	0.00	92.41	-0.26
48.000	0.00	92.41	-0.05
51.000	-5811.62	92.41	89.46
54.000	-5811.62	-17.76	89.46
57.000	-5811.62	-3.35	89.46
60.000	-5811.62	-0.63	89.46
63.000	-5811.62	-0.12	89.46
66.000	-5811.62	-0.02	89.46
69.000	-5811.62	-5812.20	89.46
72.000	-16.31	-5812.20	-8.17
75.000	-3.08	-5812.20	-1.54
78.000	-0.58	-5812.20	-0.29
81.000	-0.11	-5812.20	-0.06
84.001	-5812.79	-5812.20	-86.71
87.001	-5812.78	-39.67	-86.71
90.001	-5812.78	-7.50	-86.71
93.001	-5812.78	-1.42	-86.71
96.001	-5812.78	-0.27	-86.71
99.001	-5812.78	-0.05	-86.71
102.001	-5812.78	-89.60	-86.71
105.001	-18.24	-89.60	-10.50
108.001	-3.45	-89.60	-7.72
111.001	-0.65	-89.60	-2.54
114.001	-0.12	-89.60	-0.69
117.001	5629.73	-89.60	-86.64
120.001	5629.65	13.89	-86.66
123.001	5629.65	-11.34	-86.66
126.001	5629.65	-4.78	-86.66
129.001	5629.65	-1.40	-86.66
132.001	5629.65	-0.36	-86.66
135.001	5629.65	5630.28	-86.66
138.001	1.42	5630.28	-9.45
141.001	-12.55	5630.28	-8.21
144.001	-4.79	5630.28	-2.77
147.001	-1.36	5630.28	-0.75
150.001	5631.79	5630.28	83.74

13. TRANSIENT STABILITY (CHAPTER 13)

This program computes machine power angles and frequencies in a three-phase power system subjected to disturbances. The program also computes machine angular accelerations, machine electrical power outputs, and bus voltage magnitudes.

Before running this program, it is first necessary to run CHAPTER 7 POWER FLOW, for your power system under consideration. CHAPTER 13 opens and then reads data from the CHAPTER 7 input/output data files.

Input data for CHAPTER 13 also include synchronous machine data, which are stored in a data file. The user can store up to three separate cases. The synchronous machine data file for Case 1 is called TSM1.DAT. The data file for Case 2 (3) has a similar name, except the number 2 (3) is used. The maximum number of machines is 25.

Synchronous machine data include the machine bus number, the per-unit transient reactance, and the per-unit H constant for each machine.

The user has the following options: (1) update the synchronous machine input data; (2) set the disturbances and run the program; (3) stop.

When the user updates the synchronous machine input data, the following options are given: (1) initialize the file (which removes all data currently stored); (2) add a new machine or change an existing one; (3) remove a machine; (4) display the data for one machine; and (5) display the data for all machines after a given one.

When the user sets the disturbances and runs the program, the integration interval Δt (s) and final time t_F (s) are first selected. Δt should be small in order that an accurate solution is obtained. Typical values are $\Delta t = 0.01$ s or $\Delta t = 0.01667$ s (one cycle in a 60 Hz system).

Next the user selects the number of disturbances (1, 2, 3, or 4) and the disturbance times (s). The user has the following options for the first disturbance: (1) put a three-phase short circuit at a bus; and (2) open one or more machine breakers, transmission line breakers, or transformer breakers.

The user has the following disturbance options after the first disturbance: (1) put a three-phase short circuit at a bus; (2) extinguish an existing short circuit; (3) open one or more machine breakers, transmission line breakers, or transformer breakers; and (4) close one or more machine breakers, transmission line breakers, or transformer breakers.

The program alternately solves, step-by-step,. algebraic power flow equations representing the network and differential equations representing the machines. The Gauss-Seidel iterative method is used to solve the power flow equations, and a modified Euler's method is used to solve the differential equations.

The user can select from the following outputs: (1) machine power angle; (2) machine frequency; (3) machine real power output; and (4) bus voltage. Up to five outputs can be displayed during any run. The user also selects the printout integer K. Then the outputs are displayed at intervals of KΔt. The outputs can be displayed in exponential format (scientific notation) or in fixed-point format (numbers with decimal points).

The following sample run is for a temporary three-phase short circuit on line 6 at bus 5 of the power system whose single-line diagram is shown in Figure 13a. Note that this is the same single-line diagram as that in Figure 7b. For prefault conditions, the power flow output data from the CHAPTER 7 sample run are used. Machine input data are given in the CHAPTER 13 sample run. The short circuit is cleared by opening circuit breakers B42 and B52 at t=0.05 s (3 cycles), followed by reclosing these breakers at t=0.50 s (30 cycles after fault clearing). When reclosure occurs, the temporary fault has already been extinguished. the integration interval is Δt=0.01 s and the final time is t_F=0.75 s.

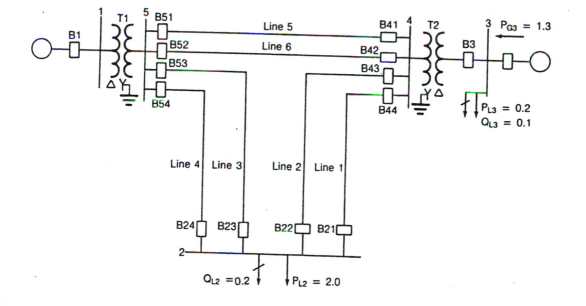

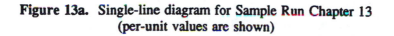

**Figure 13a. Single-line diagram for Sample Run Chapter 13
(per-unit values are shown)**

SAMPLE RUN CHAPTER 13 FIVE-BUS SYSTEM

WHAT IS YOUR POWER-FLOW CASE NUMBER (ENTER AN INTEGER FROM 1 TO 3) 1

ENTER YOUR CASE NAME SAMPLE RUN

WOULD YOU LIKE TO:
```
            1. UPDATE THE SYNCHRONOUS MACHINE INPUT DATA
            2. SET DISTURBANCES AND RUN THE PROGRAM
            3. STOP
```

ENTER YOUR SELECTION (1,2 OR 3) 1

SELECTIONS:
```
 1. INITIALIZE FILE
 2. ADD A NEW SYNCHRONOUS MACHINE  OR CHANGE AN EXISTING   ONE
 3. REMOVE A MACHINE
 4. DISPLAY ONE MACHINE
 5. DISPLAY ALL MACHINES BELOW A GIVEN ONE
 6. DONE WITH MACHINE DATA
```

ENTER YOUR SELECTION (1 TO 6) 5
ENTER THE NUMBER OF THE FIRST MACHINE TO BE DISPLAYED 1

USE THE Ctrl PRINT SCREEN OPTION NOW IF YOU WANT TO PRINT THE RESULTS.

PRESS RETURN TO CONTINUE.
            SYNCHRONOUS MACHINE DATA FOR SAMPLE RUN

MACHINE#	BUS	X' per unit	H per unit seconds
1	1	.2	5
2	3	.1	50

REMOVE Ctrl PRINT SCREEN AND THEN PRESS RETURN TO CONTINUE

SELECTIONS:
```
 1. INITIALIZE FILE
 2. ADD A NEW SYNCHRONOUS MACHINE  OR CHANGE AN EXISTING   ONE
 3. REMOVE A MACHINE
 4. DISPLAY ONE MACHINE
 5. DISPLAY ALL MACHINES BELOW A GIVEN ONE
 6. DONE WITH MACHINE DATA
```

ENTER YOUR SELECTION (1 TO 6) 6

WOULD YOU LIKE TO:
```
            1. UPDATE THE SYNCHRONOUS MACHINE INPUT DATA
            2. SET DISTURBANCES AND RUN THE PROGRAM
            3. STOP
```

ENTER YOUR SELECTION (1,2 OR 3) 2

```
ENTER THE INTEGRATION TIME INTERVAL (SECONDS)  0.01
ENTER THE FINAL TIME (SECONDS)  0.760
ENTER THE NUMBER OF DISTURBANCE TIMES (1,2,3 OR 4) 3




FIRST DISTURBANCE
ENTER THE FIRST DISTURBANCE TIME(SECONDS) 0
IS THERE A SHORT CIRCUIT AT THE FIRST DISTURBANCE TIME(Y OR N) ? Y
ENTER THE SHORT CIRCUIT BUS NUMBER 5
DO GENERATOR BREAKERS OPEN AT THE FIRST DISTURBANCE(Y OR N) ? N
DO LINE   BREAKERS OPEN AT THE FIRST DISTURBANCE(Y OR N) ? N
DO TRANSFORMER  BREAKERS OPEN AT THE FIRST DISTURBANCE(Y OR N) ? N

SECOND DISTURBANCE
ENTER THE SECOND DISTURBANCE TIME(SECONDS) 0.05
IS THERE A SHORT CIRCUIT AT THE SECOND DISTURBANCE TIME(Y OR N) ? N
IS THE SHORT CIRCUIT AT THE FIRST DISTURBANCE EXTINGUISHED (Y OR N) ? Y
DO  GENERATOR  BREAKERS OPEN AT THE SECOND DISTURBANCE(Y OR N) ? N
DO LINE   BREAKERS OPEN AT THE SECOND DISTURBANCE(Y OR N) ? Y
HOW MANY LINE  BREAKERS OPEN ? 2
TRANSMISSION LINE OPENING CIRCUIT BREAKER 1
ENTER THE LINE NUMBER 6
ENTER THE BUS NUMBER 4
TRANSMISSION LINE OPENING CIRCUIT BREAKER 2
ENTER THE LINE NUMBER 6
ENTER THE BUS NUMBER 5
DO  TRANSFORMER  BREAKERS OPEN AT THE SECOND DISTURBANCE(Y OR N) ? N

THIRD DISTURBANCE
ENTER THE THIRD DISTURBANCE TIME(SECONDS) 0.50
IS THERE A SHORT CIRCUIT AT THE THIRD DISTURBANCE TIME(Y OR N) ? N
DO  GENERATOR  BREAKERS OPEN AT THE THIRD DISTURBANCE(Y OR N) ? N
DO  GENERATOR  BREAKERS RECLOSE AT THE THIRD DISTURBANCE(Y OR N) ? N
DO LINE   BREAKERS OPEN AT THE THIRD DISTURBANCE(Y OR N) ? N
DO  LINE   BREAKERS RECLOSE AT THE THIRD DISTURBANCE(Y OR N) ? Y
HOW MANY  LINE  BREAKERS CLOSE  ? 2
TRANSMISSION LINE CLOSING CIRCUIT BREAKER 1
ENTER THE LINE NUMBER 6
ENTER THE BUS NUMBER 4
TRANSMISSION LINE CLOSING CIRCUIT BREAKER 2
ENTER THE LINE NUMBER 6
ENTER THE BUS NUMBER 5
DO  TRANSFORMER  BREAKERS OPEN AT THE THIRD  DISTURBANCE(Y OR N) ? N
DO  TRANSFORMER  BREAKERS RECLOSE AT THE THIRD DISTURBANCE(Y OR N) ? N

FOR OUTPUT LABELLING
ENTER THE FIRST DISTURBANCE NAME(ANY ALPHANUMERIC SEQUENCE) FAULT AT BUS 5

ENTER THE SECOND DISTURBANCE NAME FAULT CLEARED
ENTER THE THIRD DISTURBANCE NAME                  RECLOSURE


YOU MAY PRINT FROM 1 TO 5 OUTPUTS
ENTER THE NUMBER OF OUTPUTS ( 1 TO 5 ) 5
```

```
OUTPUT  1
    SELECTIONS:
              1. MACHINE POWER ANGLE
              2. MACHINE FREQUENCY
              3. MACHINE REAL POWER OUTPUT
              4. BUS VOLTAGE

ENTER YOUR SELECTION (1,2,3 OR 4) 1
ENTER THE MACHINE NUMBER 1

OUTPUT  2
    SELECTIONS:
              1. MACHINE POWER ANGLE
              2. MACHINE FREQUENCY
              3. MACHINE REAL POWER OUTPUT
              4. BUS VOLTAGE

ENTER YOUR SELECTION (1,2,3 OR 4) 2
ENTER THE MACHINE NUMBER 1

OUTPUT  3
    SELECTIONS:
              1. MACHINE POWER ANGLE
              2. MACHINE FREQUENCY
              3. MACHINE REAL POWER OUTPUT
              4. BUS VOLTAGE

ENTER YOUR SELECTION (1,2,3 OR 4) 1
ENTER THE MACHINE NUMBER 2

OUTPUT  4
    SELECTIONS:
              1. MACHINE POWER ANGLE
              2. MACHINE FREQUENCY
              3. MACHINE REAL POWER OUTPUT
              4. BUS VOLTAGE

ENTER YOUR SELECTION (1,2,3 OR 4) 2
ENTER THE MACHINE NUMBER 2

OUTPUT  5
    SELECTIONS:
              1. MACHINE POWER ANGLE
              2. MACHINE FREQUENCY
              3. MACHINE REAL POWER OUTPUT
              4. BUS VOLTAGE

ENTER YOUR SELECTION (1,2,3 OR 4) 4
ENTER THE BUS NUMBER 5
THE OUTPUTS ARE PRINTED EVERY KTH  TIME INTERVAL
ENTER THE PRINTOUT INTEGER K  ( K >= 1 ) 3
DO YOU WANT THE OUTPUTS IN EXPONENTIAL FORMAT (Y OR N) ? N
CONTINUE ( Y ) OR RESET THE DISTURBANCES (N) ? Y
USE THE Ctrl PRINT SCREEN OPTION NOW IF YOU WANT TO PRINT THE OUTPUTS
PRESS RETURN TO CONTINUE
```

TRANSIENT STABILITY OUTPUT FOR   SAMPLE RUN

TIME	DELTA 1	OMEGA 1	DELTA 2	OMEGA 2	V 5
seconds	degrees	rad/s	degrees	rad/s	per unit

FAULT AT BUS 5

TIME	DELTA 1	OMEGA 1	DELTA 2	OMEGA 2	V 5
0.000	12.72	376.99	5.77	376.99	0.000
0.030	13.58	377.99	5.86	377.11	0.000

FAULT CLEARED

TIME	DELTA 1	OMEGA 1	DELTA 2	OMEGA 2	V 5
0.060	16.14	378.98	6.16	377.22	0.999
0.090	19.45	378.83	6.58	377.24	0.996
0.120	22.41	378.58	7.04	377.28	0.994
0.150	24.87	378.25	7.56	377.32	0.991
0.180	26.73	377.87	8.16	377.36	0.989
0.210	27.90	377.47	8.84	377.41	0.988
0.240	28.36	377.06	9.61	377.46	0.989
0.270	28.13	376.67	10.47	377.51	0.990
0.300	27.28	376.33	11.40	377.55	0.992
0.330	25.89	376.06	12.40	377.59	0.995
0.360	24.12	375.88	13.45	377.61	0.998
0.390	22.14	375.82	14.53	377.63	1.001
0.420	20.14	375.86	15.63	377.63	1.003
0.450	18.30	376.01	16.72	377.62	1.004
0.480	16.82	376.26	17.79	377.60	1.004

RECLOSURE

TIME	DELTA 1	OMEGA 1	DELTA 2	OMEGA 2	V 5
0.510	15.84	376.60	18.82	377.57	1.022
0.540	15.52	377.02	19.78	377.53	1.024
0.570	15.96	377.48	20.66	377.48	1.024
0.600	17.19	377.93	21.45	377.43	1.024
0.630	19.18	378.35	22.17	377.38	1.024
0.660	21.83	378.70	22.81	377.35	1.024
0.690	25.01	378.96	23.39	377.32	1.024
0.720	28.53	379.10	23.93	377.30	1.023
0.750	32.20	379.13	24.45	377.29	1.021

REMOVE Ctrl PRINT SCREEN AND THEN PRESS RETURN TO CONTINUE

WOULD YOU LIKE TO:
         1. UPDATE THE SYNCHRONOUS MACHINE INPUT DATA
         2. SET DISTURBANCES AND RUN THE PROGRAM
         3. STOP

ENTER YOUR SELECTION (1,2 OR 3) 3

14. PROGRAM LISTINGS AND FLOW CHARTS

This section is primarily intended for those readers who are interested in writing their own computer programs. Program listings of the nine subroutines in the program CHAPTER 2 and flow charts for the remaining programs are given.

Background and theory for programs are given in the text *Power System Analysis and Design with Personal Computer Applications*. Equation numbers from the text are listed in the flow charts given here.

```
11000 '**********SUBROUTINE RMA(A,B,C,N,M)***********
11010 'THIS SUBROUTINE COMPUTES THE MATRIX SUM C = A + B
11020 'OF THE TWO NxM MATRICES A AND B
11030 '***************************************************
11040 FOR I=1 TO N
11050 FOR J=1 TO M
11060 C(I,J)=A(I,J)+B(I,J)
11070 NEXT J
11080 NEXT I
11090 RETURN
11100 '************END OF RMA SUBROUTINE******************
12000 '********SUBROUTINE CMA(A,B,C,N,M)***************
12010 'THIS SUBROUTINE COMPUTES THE MATRIX SUM C = A + B
12020 ' OF THE TWO NxM COMPLEX MATRICES A AND B
12030 '***************************************************
12040 FOR I=1 TO N
12050 FOR J=1 TO M
12060 CR(I,J)=AR(I,J)+BR(I,J)
12070 CI(I,J)=AI(I,J)+BI(I,J)
12080 NEXT J
12090 NEXT I
12100 RETURN
12110 '************END OF CMA SUBROUTINE******************
13000 '**********SUBROUTINE RMM(A,B,C,N,M,P)***********
13010 'THIS SUBROUTINE COMPUTES THE MATRIC PRODUCT C = AB
13020 'OF THE NxM REAL MATRIX A AND THE MxP REAL MATRIX B.
13030 'THE RESULTING MATRIX C HAS DIMENSION NxP
13040 '***************************************************
13050 FOR I=1 TO N
13060 FOR J=1 TO P
13070 C(I,J)=0
13080 NEXT J
13090 NEXT I
13100 FOR I=1 TO N
13110 FOR J=1 TO P
13120 FOR K=1 TO M
13130 C(I,J)=C(I,J)+A(I,K)*B(K,J)
13140 NEXT K
13150 NEXT J
13160 NEXT I
13170 RETURN
13180 '**********END OF RMM SUBROUTINE*******************
```

```
14000 '**************SUBROUTINE CMM(A,B,C,N,M,P)***********
14010 'THIS SUBROUTINE COMPUTES THE MATRIX PRODUCT C = A B
14020 'OF THE TWO NxM COMPLEX MATRICES A AND B
14030 'THE RESULTING COMPLEX MATRIX C HAS DIMENSION NxP
14040 '**************************************************
14050 FOR I=1 TO N
14060 FOR J=1 TO P
14070 CR(I,J)=0
14080 CI(I,J)=0
14090 NEXT J
14100 NEXT I
14110 FOR I=1 TO N
14120 FOR J=1 TO P
14130 FOR K=1 TO M
14140 CR(I,J)=CR(I,J)+AR(I,K)*BR(K,J)-AI(I,K)*BI(K,J)
14150 CI(I,J)=CI(I,J)+AR(I,K)*BI(K,J)+AI(I,K)*BR(K,J)
14160 NEXT K
14170 NEXT J
14180 NEXT I
14190 RETURN
14200 '***********END OF CMM SUBROUTINE*****************


15000 '***********SUBROUTINE RMI(A,N)********************
15010 'THIS SUBROUTINE COMPUTES THE INVERSE OF
15020 'THE NxN REAL MATRIX A,WHOSE DETERMINANT IS
15030 'ASSUMED TO BE NONZERO.
15040 'THE GAUSS ELIMINATION METHOD IS EMPLOYED.
15050 'IT IS ALSO ASSUMED THAT THERE IS NO ZERO PIVOT ELEMENT.
15060 'THE INVERSE MATRIX IS CALLED A AND IS STORED IN THE SAME
15070 'LOCATION AS THE ORIGINAL MATRIX. THE ORIGINAL A MATRIX
15080 'IS ERASED.
15090 FOR I=1 TO N
15100 IF A(I,I)<>0 THEN GOTO 15130
15110 PRINT " THE GAUSS ELIMINATION METHOD GIVES A ZERO PIVOT ELEMENT.
15112 PRINT " PLEASE CHANGE YOUR MATRIX."
15120 GOTO 5230
15130 A(I,I)=1/A(I,I)
15140 FOR J=1 TO N
15150 IF J=I GOTO 15230
15160 A(J,I)=A(J,I)*A(I,I)
15170 FOR K=1 TO N
15180 IF K=I GOTO 15220
15190 A(J,K)=A(J,K)-A(J,I)*A(I,K)
15200 IF J=N THEN 15210 ELSE 15220
15210 A(I,K)=-A(I,I)*A(I,K)
15220 NEXT K
15230 NEXT J
15240 NEXT I
15250 K=N-1
15260 FOR J=1 TO K
15270 A(N,J)=-A(N,N)*A(N,J)
15280 NEXT J
15290 RETURN
15300 '***********END OF RMI SUBROUTINE*****************
```

```
16000 '********SUBROUTINE CMI(A,N)***************************
16010 'THIS SUBROUTINE COMPUTES THE INVERSE OF
16020 'THE NxN COMPLEX MATRIX A = AR +jAI, WHOSE DETERMINANT IS
16030 'ASSUMED TO BE NONZERO.
16040 'THE GAUSS ELIMINATION METHOD IS USED.
16050 'IT IS ALSO ASSUMED THAT THERE IS NO ZERO PIVOT ELEMENT.
16060 'THE INVERSE MATRIX IS CALLED A AND IS STORED IN THE
16070 'LOCATION AS THE ORIGINAL MATRIX. THE ORIGINAL
16080 'A MATRIX IS THEREFORE ERASED.
16090 FOR I=1 TO N
16100 ASQ=AR(I,I)^2+AI(I,I)^2
16110 IF ASQ<>0 THEN GOTO 16140
16120 PRINT " THE GAUSS ELIMINATION METHOD GIVES A ZERO PIVOT ELEMENT."
16122 PRINT " PLEASE CHANGE YOUR MATRIX."
16130 GOTO 6560
16140 AR(I,I)=AR(I,I)/ASQ
16150 AI(I,I)=-AI(I,I)/ASQ
16160 FOR J=1 TO N
16170 IF J=I GOTO 16340
16180 BBB=AR(J,I)*AR(I,I)-AI(J,I)*AI(I,I)
16190 CCC=AR(J,I)*AI(I,I)+AI(J,I)*AR(I,I)
16200 AR(J,I)=BBB
16210 AI(J,I)=CCC
16220 FOR K=1 TO N
16230 IF K=I GOTO 16330
16240 DDD=AR(J,K)-AR(J,I)*AR(I,K)+AI(J,I)*AI(I,K)
16250 EEE=AI(J,K)-AR(J,I)*AI(I,K)-AI(J,I)*AR(I,K)
16260 AR(J,K)=DDD
16270 AI(J,K)=EEE
16280 IF J=N THEN 16290 ELSE 16330
16290 FFF=-AR(I,I)*AR(I,K)+AI(I,I)*AI(I,K)
16300 GGG=-AR(I,I)*AI(I,K)-AI(I,I)*AR(I,K)
16310 AR(I,K)=FFF
16320 AI(I,K)=GGG
16330 NEXT K
16340 NEXT J
16350 NEXT I
16360 K=N-1
16370 FOR J=1 TO K
16380 HHH=-AR(N,N)*AR(N,J)+AI(N,N)*AI(N,J)
16390 PPP=-AR(N,N)*AI(N,J)-AI(N,N)*AR(N,J)
16400 AR(N,J)=HHH
16410 AI(N,J)=PPP
16420 NEXT J
16430 RETURN
16440 '***********END OF CMI SUBROUTINE********************
```

```
17000 '*********SUBROUTINE RMT(A,AT,N,M)*******************
17010 'THIS SUBROUTINE COMPUTES THE MATRIX TRANSPOSE A T
17015 ' OF THE REAL MATRIX A.
17020 FOR I=1 TO M
17030 FOR J=1 TO N
17040 AT(I,J)=A(J,I)
17050 NEXT J
17060 NEXT I
17070 RETURN
17080 '*******END OF RMT SUBROUTINE*************************


18000 '*********SUBROUTINE CMT(A,AT,N,M)*********************
18010 'THIS SUBROUTINE COMPUTES THE MATRIX TRANSPOSE AT
18020 'OF THE COMPLEX MATRIX A.
18030 FOR I=1 TO M
18040 FOR J=1 TO N
18050 ATR(I,J)=AR(J,I)
18060 ATI(I,J)=AI(J,I)
18070 NEXT J
18080 NEXT I
18090 RETURN
18100 '**************END OF CMT SUBROUTINE*********************


19000 '**********SUBROUTINE CMC(A,AC,N,M)*********************
19010 'THIS SUBROUTINE COMPUTES THE MATRIX CONJUGATE AC
19020 'OF THE COMPLEX MATRIX A.
19030 FOR I=1 TO N
19040 FOR J=1 TO M
19050 ACR(I,J)=AR(I,J)
19060 ACI(I,J)=-AI(I,J)
19070 NEXT J
19080 NEXT I
19090 RETURN
19100 '*******END OF CMC SUBROUTINE*************************
```

CHAPTER 3 SYMMETRICAL COMPONENTS

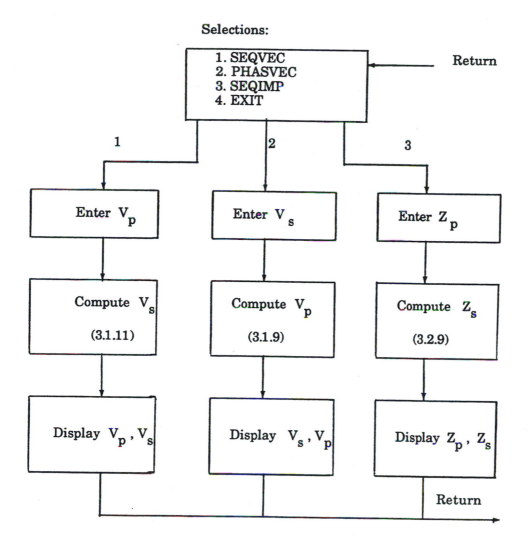

Selections:

1. SEQVEC
2. PHASVEC
3. SEQIMP
4. EXIT

Return

1

2

3

Enter V_p

Enter V_s

Enter Z_p

Compute V_s

(3.1.11)

Compute V_p

(3.1.9)

Compute Z_s

(3.2.9)

Display V_p, V_s

Display V_s, V_p

Display Z_p, Z_s

Return

CHAPTER 5 LINE CONSTANTS

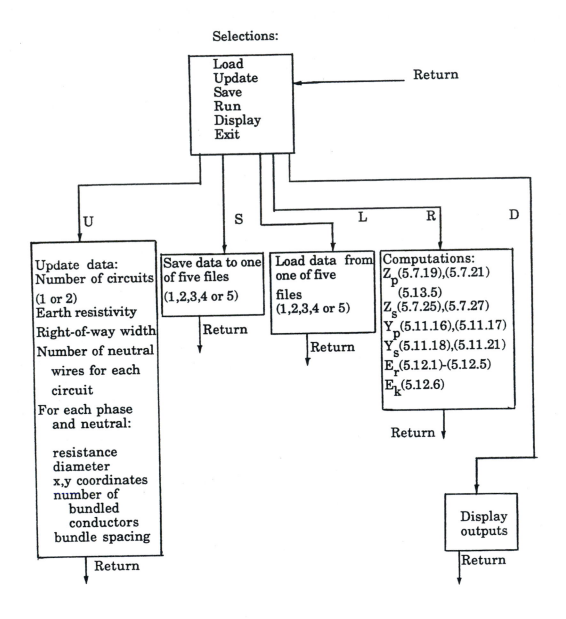

Selections:

Load
Update
Save
Run
Display
Exit

Return

U   S   L   R   D

Update data:
Number of circuits
(1 or 2)
Earth resistivity
Right-of-way width
Number of neutral
    wires for each
    circuit
For each phase
and neutral:

    resistance
    diameter
    x,y coordinates
    number of
        bundled
        conductors
    bundle spacing

Return

Save data to one
of five files
(1,2,3,4 or 5)

Return

Load data from
one of five
files
(1,2,3,4 or 5)

Return

Computations:
Z_p(5.7.19),(5.7.21)
    (5.13.5)
Z_s(5.7.25),(5.7.27)
Y_p(5.11.16),(5.11.17)
Y_s(5.11.18),(5.11.21)
E_r(5.12.1)-(5.12.5)
E_k(5.12.6)

Return

Display
outputs

Return

CHAPTER 6 TRANSMISSION LINES - STEADY STATE

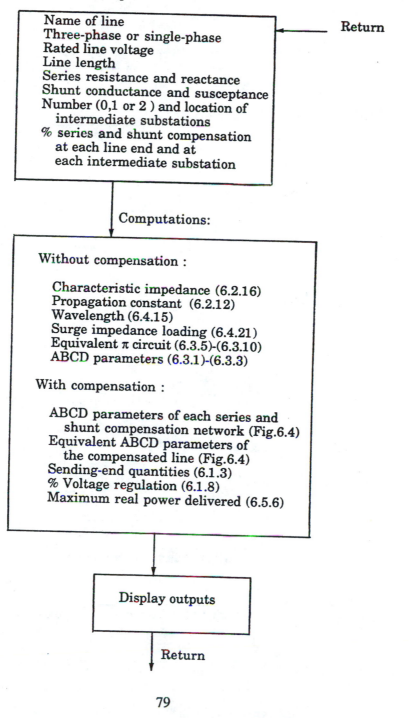

Enter the input data:

Name of line
Three-phase or single-phase
Rated line voltage
Line length
Series resistance and reactance
Shunt conductance and susceptance
Number (0,1 or 2) and location of
    intermediate substations
% series and shunt compensation
    at each line end and at
    each intermediate substation

Return

Computations:

Without compensation :

Characteristic impedance (6.2.16)
Propagation constant (6.2.12)
Wavelength (6.4.15)
Surge impedance loading (6.4.21)
Equivalent π circuit (6.3.5)-(6.3.10)
ABCD parameters (6.3.1)-(6.3.3)

With compensation :

ABCD parameters of each series and
    shunt compensation network (Fig.6.4)
Equivalent ABCD parameters of
    the compensated line (Fig.6.4)
Sending-end quantities (6.1.3)
% Voltage regulation (6.1.8)
Maximum real power delivered (6.5.6)

Display outputs

Return

CHAPTER 7 POWER FLOW

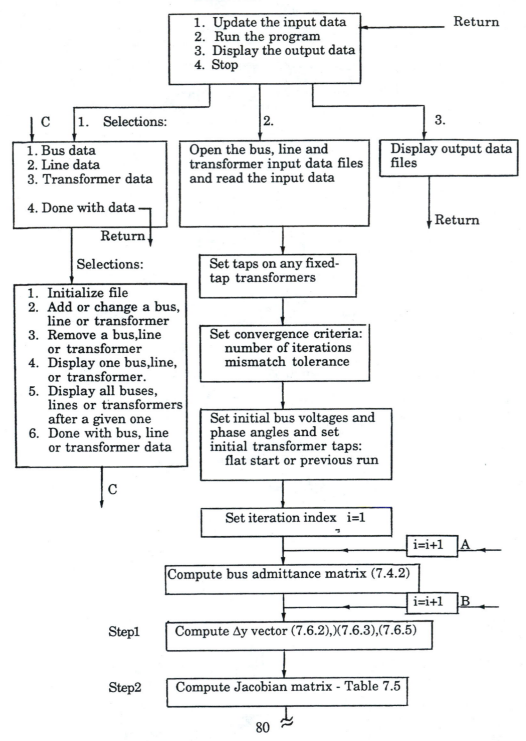

Selections:

1. Update the input data
2. Run the program
3. Display the output data
4. Stop

Return

C 1. Selections:

1. Bus data
2. Line data
3. Transformer data
4. Done with data

Return

Selections:

1. Initialize file
2. Add or change a bus, line or transformer
3. Remove a bus, line or transformer
4. Display one bus, line, or transformer.
5. Display all buses, lines or transformers after a given one
6. Done with bus, line or transformer data

C

2.

Open the bus, line and transformer input data files and read the input data

Set taps on any fixed-tap transformers

Set convergence criteria: number of iterations mismatch tolerance

Set initial bus voltages and phase angles and set initial transformer taps: flat start or previous run

Set iteration index i=1

i=i+1 A

Compute bus admittance matrix (7.4.2)

i=i+1 B

Step1 Compute Δy vector (7.6.2),)(7.6.3)(7.6.5)

Step2 Compute Jacobian matrix - Table 7.5

3.

Display output data files

Return

80 ≈

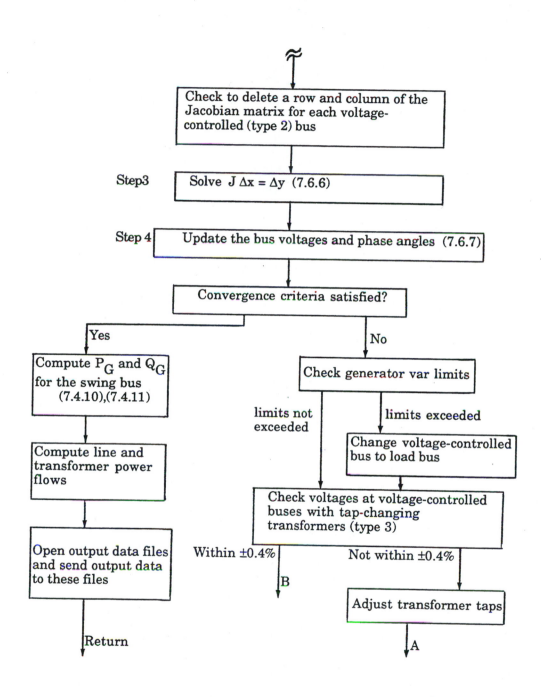

Check to delete a row and column of the Jacobian matrix for each voltage-controlled (type 2) bus

Step3 Solve $J \Delta x = \Delta y$ (7.6.6)

Step 4 Update the bus voltages and phase angles (7.6.7)

Convergence criteria satisfied?

Yes

Compute P_G and Q_G for the swing bus (7.4.10),(7.4.11)

Compute line and transformer power flows

Open output data files and send output data to these files

Return

No

Check generator var limits

limits not exceeded

limits exceeded

Change voltage-controlled bus to load bus

Check voltages at voltage-controlled buses with tap-changing transformers (type 3)

Within ±0.4%

B

Not within ±0.4%

Adjust transformer taps

A

CHAPTER 8 SYMMETRICAL SHORT CIRCUITS
(CHAPTER 9 SHORT CIRCUITS)

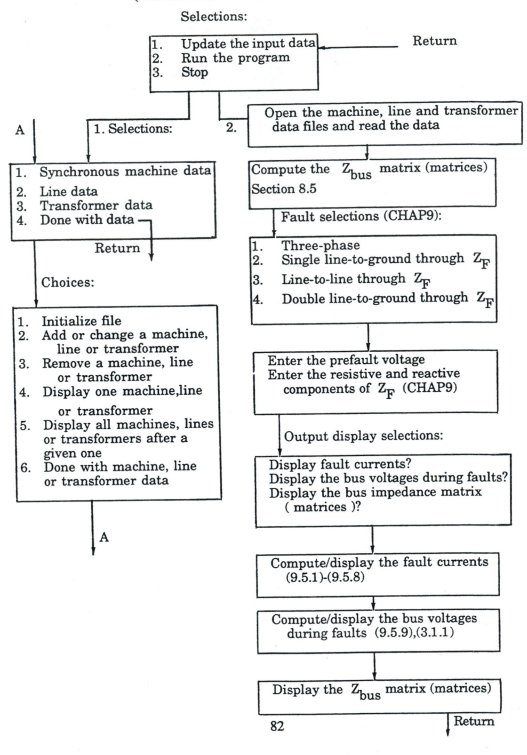

Selections:

1.	Update the input data
2.	Run the program
3.	Stop

Return

A   1. Selections: 2.   Open the machine, line and transformer data files and read the data

1.	Synchronous machine data
2.	Line data
3.	Transformer data
4.	Done with data

Compute the Z_{bus} matrix (matrices)

Section 8.5

Return

Fault selections (CHAP9):

Choices:

1.	Three-phase
2.	Single line-to-ground through Z_F
3.	Line-to-line through Z_F
4.	Double line-to-ground through Z_F

1.	Initialize file
2.	Add or change a machine, line or transformer
3.	Remove a machine, line or transformer
4.	Display one machine,line or transformer
5.	Display all machines, lines or transformers after a given one
6.	Done with machine, line or transformer data

Enter the prefault voltage
Enter the resistive and reactive components of Z_F (CHAP9)

Output display selections:

Display fault currents?
Display the bus voltages during faults?
Display the bus impedance matrix (matrices)?

A

Compute/display the fault currents (9.5.1)-(9.5.8)

Compute/display the bus voltages during faults (9.5.9),(3.1.1)

Display the Z_{bus} matrix (matrices)

Return

CHAPTER 12 TRANSMISSION LINE TRANSIENTS

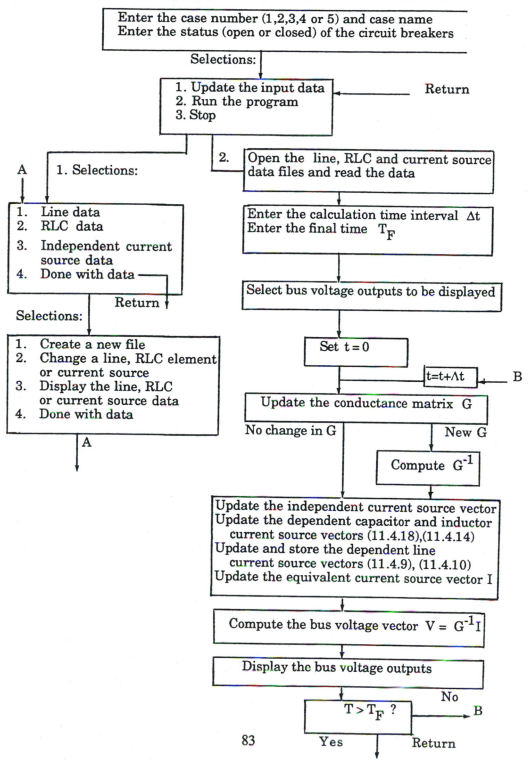

Enter the case number (1,2,3,4 or 5) and case name
Enter the status (open or closed) of the circuit breakers

Selections:

1. Update the input data
2. Run the program
3. Stop

Return

A   1. Selections:

2.   Open the line, RLC and current source data files and read the data

1.   Line data
2.   RLC data
3.   Independent current source data
4.   Done with data

Return

Enter the calculation time interval Δt
Enter the final time T_F

Selections:

1.   Create a new file
2.   Change a line, RLC element or current source
3.   Display the line, RLC or current source data
4.   Done with data

Select bus voltage outputs to be displayed

Set $t = 0$

$t = t + \Delta t$ B

A

Update the conductance matrix G

No change in G   New G

Compute G^{-1}

Update the independent current source vector
Update the dependent capacitor and inductor
    current source vectors (11.4.18),(11.4.14)
Update and store the dependent line
    current source vectors (11.4.9), (11.4.10)
Update the equivalent current source vector I

Compute the bus voltage vector $V = G^{-1}I$

Display the bus voltage outputs

No

$T > T_F$?   B

83   Yes   Return

CHAPTER 13 TRANSIENT STABILITY

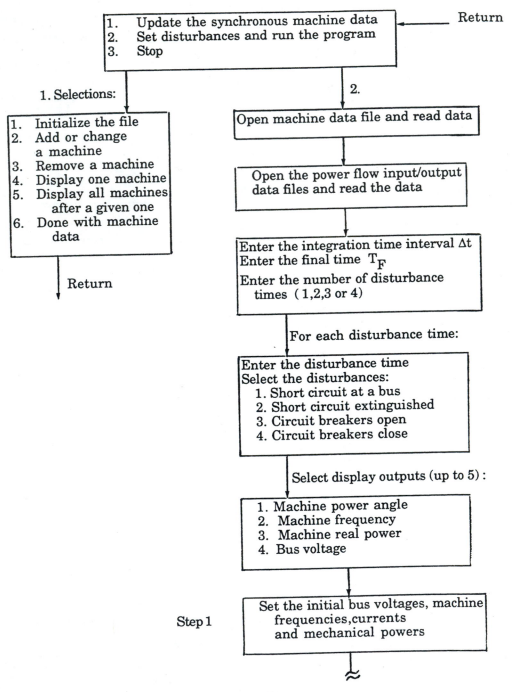

Selections:

1.	Update the synchronous machine data
2.	Set disturbances and run the program
3.	Stop

Return

1. Selections:

1.	Initialize the file
2.	Add or change a machine
3.	Remove a machine
4.	Display one machine
5.	Display all machines after a given one
6.	Done with machine data

Return

2.

Open machine data file and read data

Open the power flow input/output data files and read the data

Enter the integration time interval Δt
Enter the final time T_F
Enter the number of disturbance times (1,2,3 or 4)

For each disturbance time:

Enter the disturbance time
Select the disturbances:
 1. Short circuit at a bus
 2. Short circuit extinguished
 3. Circuit breakers open
 4. Circuit breakers close

Select display outputs (up to 5) :

1. Machine power angle
2. Machine frequency
3. Machine real power
4. Bus voltage

Step 1

Set the initial bus voltages, machine frequencies,currents and mechanical powers

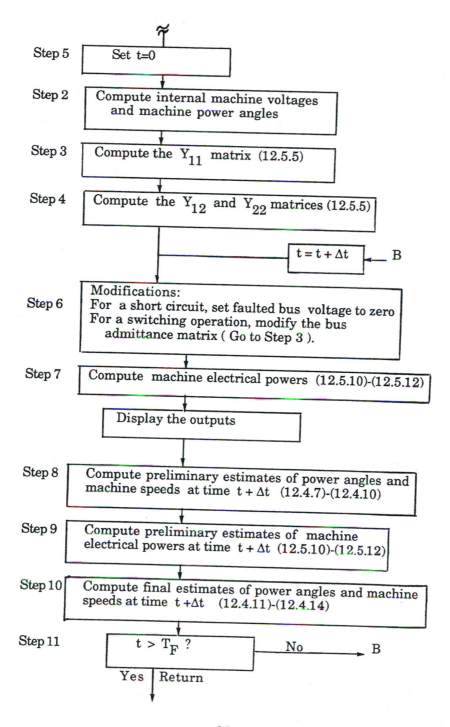

Step 5 — Set t=0

Step 2 — Compute internal machine voltages and machine power angles

Step 3 — Compute the Y_{11} matrix (12.5.5)

Step 4 — Compute the Y_{12} and Y_{22} matrices (12.5.5)

$t = t + \Delta t$ ◄— B

Step 6 — Modifications:
For a short circuit, set faulted bus voltage to zero
For a switching operation, modify the bus
admittance matrix (Go to Step 3).

Step 7 — Compute machine electrical powers (12.5.10)-(12.5.12)

Display the outputs

Step 8 — Compute preliminary estimates of power angles and machine speeds at time $t + \Delta t$ (12.4.7)-(12.4.10)

Step 9 — Compute preliminary estimates of machine electrical powers at time $t + \Delta t$ (12.5.10)-(12.5.12)

Step 10 — Compute final estimates of power angles and machine speeds at time $t + \Delta t$ (12.4.11)-(12.4.14)

Step 11 — $t > T_F$?   No   B

Yes | Return

85

15. SAMPLE SINGLE-LINE DIAGRAMS

This section contains single-line diagrams and input data for original versions of IEEE 14-bus, 30-bus and 57-bus test systems, which are excerpted from the following publication: L.L. Freris and A.M. Sasson, "Investigation of the Load-Flow Problem," *IEEE Proceedings*, Vol 115, No. 10, pp. 1459-1470, October, 1968.

Included in these single-line diagrams are synchronous condensers, three-winding transformers and static capacitors, which are briefly discussed in the following paragraphs.

SYNCHRONOUS CONDENSERS

A synchronous condenser, which is a synchronous motor operating at no-load, delivers or absorbs a variable amount of reactive power. In the power-flow program, the bus to which the condenser is connected is represented as a voltage-controlled (type 2) bus with zero power generation. Synchronous condensers are identified by the letter C in the enclosed single-line diagrams.

THREE-WINDING TRANSFORMERS

A three-winding transformer can be represented by three two-winding transformers connected in a T (or Y) circuit, as described in Section 4.6 of the text, *Power System Analysis and Design with Personal Computer Applications*. The equivalent impedances of the T circuit, computed from the transformer leakage reactances, are listed in the transmission line and transformer tables here.

STATIC CAPACITORS

A static capacitor, which is a capacitor bank connected from bus to ground, delivers reactive power. In the power-flow program, it can be represented by a load consisting of zero real power and negative reactive power $Q_L = -(V^2)(B_C)$ per unit, where V is the per-unit bus voltage and B_C is the per-unit susceptance of the static capacitor. At bus voltages near 1.0 per unit, this can be approximated by a constant negative reactive load $Q_L \approx -B_C$.

IEEE 14-BUS TEST SYSTEM

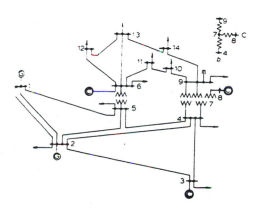

Figure 13a Single-line diagram

Line designation	Resistance p.u.*	Reactance p.u.*	Line charging p.u.*
1–2	0·01938	0·05917	0·0264
1–5	0·05403	0·22304	0·0246
2–3	0·04699	0·19797	0·0219
2–4	0·05811	0·17632	0·0187
2·5	0·05695	0·17388	0·0170
3–4	0·06701	0·17103	0·0173
4–5	0·01335	0·04211	0·0064
4–7	0	0·20912	0
4–9	0	0·55618	0
5–6	0	0·25202	0
6–11	0·09498	0·19890	0
6–12	0·12291	0·25581	0
6–13	0·06615	0·13027	0
7–8	0	0·17615	0
7–9	0	0·11001	0
9–10	0·03181	0·08450	0
9–14	0·12711	0·27038	0
10–11	0·08205	0·19207	0
12–13	0·22092	0·19988	0
13–14	0·17093	0·34802	0

* Impedance and line-charging susceptance in p.u. on a 100000 kVA base
Line charging one-half of total charging of line

Table 13A Transmission line & transformer data

Bus number	Starting bus voltage		Generation		Load	
	Magnitude p.u.	Phase angle deg	MW	MVAr	MW	MVAr
1*	1·06	0	0	0	0	0
2	1·0	0	40	0	21·7	12·7
3	1·0	0	0	0	94·2	19·0
4	1·0	0	0	0	47·8	3·9
5	1·0	0	0	0	7·6	1·6
6	1·0	0	0	0	11·2	7·5
7	1·0	0	0	0	0	0
8	1·0	0	0	0	0	0
9	1·0	0	0	0	29·5	16·6
10	1·0	0	0	0	9·0	5·8
11	1·0	0	0	0	3·5	1·8
12	1·0	0	0	0	6·1	1·6
13	1·0	0	0	0	13·5	5·8
14	1·0	0	0	0	14·9	5·0

* Swing machine

Table 13B Bus data

Bus number	Voltage magnitude, p.u.	Minimum MVAr capability	Maximum MVAr capability
2	1·045	40	50
3	1·010	0	40
6	1·070	6	24
8	1·090	6	24

Table 13C Voltage-controlled (type 2) bus data

Transformer designation	Tap setting*
4–7	1.022
4–9	1.032
5–6	1.073

Table 13D Tap settings of fixed-tap transformers (The tap is on the 2nd bus number.)

Bus number	Susceptance** p.u.
9	0·19

** Susceptance in p.u. on a 100000 kVA base

Table 13 E Static capacitor data

IEEE 30-BUS TEST SYSTEM

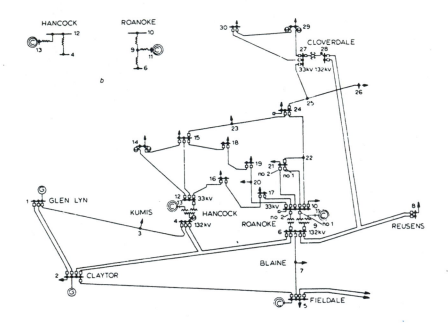

Figure 13b Single-line diagram

Bus number	Voltage magnitude p.u.	Minimum MVAr capability	Maximum MVAr capability
2	1·045	−40	50
5	1·01	−40	40
8	1·01	− 10	40
11	1·082	− 6	24
13	1·071	− 6	24

Table 13F Voltage-controlled (type 2) bus data

Transformer desgination	Tap setting*
12–4	0·932
9–6	0·978
10–6	0·969
27–28	0·968

Table 13G Tap settings of fixed-tap transformers (The tap is on the $\underline{2}^{nd}$ bus number.)

IEEE 30-BUS TEST SYSTEM

Bus number	Starting bus voltage		Generation		Load	
	Magnitude p.u.	Phase angle degrees	MW	MVAr	MW	MVAr
1*	1·06	0	0	0	0	0
2	1·0	0	40	0	21·7	12·7
3	1·0	0	0	0	2·4	1·2
4	1·0	0	0	0	7·6	1·6
5	1·0	0	0	0	94·2	19·0
6	1·0	0	0	0	0	0
7	1·0	0	0	0	22·8	10·9
8	1·0	0	0	0	30·0	30·0
9	1·0	0	0	0	0	0
10	1·0	0	0	0	5·8	2·0
11	1·0	0	0	0	0	0
12	1·0	0	0	0	11·2	7·5
13	1·0	0	0	0	0	0
14	1·0	0	0	0	6·2	1·6
15	1·0	0	0	0	8·2	2·5
16	·1·0	0	0	·0	3·5	1·8
17	1·0	0	0	0	9·0	5·8
18	1·0	0	0	0	3·2	0·9
19	1·0	0	0	0	9·5	3·4
20	1·0	0	0	0	2·2	0·7
21	1·0	0	0	0	17·5	11·2
22	1·0	0	0	0	0	0
23	1·0	0	0	0	3·2	1·6
24	1·0	0	0	0	8·7	6·7
25	1·0	0	0	0	0	0
26	1·0	0	0	0	3·5	2·3
27	1·0	0	0	0	0	0
28	1·0	0	0	0	0	0
29	1·0	0	0	0	2·4	0·9
30	1·0	0	0	0	10·6·	1·9

* Swing machine

Table 13H   Bus data

Line designation	Resistance p.u.*	Reactance p.u.*	Line charging p.u.*
1–2	0·0192	0·0575	0·0264
1–3	0·0452	0·1852	0·0204
2–4	0·0570	0·1737	0·0184
3–4	0·0132	0·0379	0·0042
2–5	0·0472	0·1983	0·0209
2–6	0·0581	0·1763	0·0187
4–6	0·0119	0·0414	0·0045
5–7	0·0460	0·1160	0·0102
6–7	0·0267	0·0820	0·0085
6–8	0·0120	0·0420	0·0045
6–9	0	0·2080	0
6–10	0	0·5560	0
9–11	0	0·2080	0
9–10	0	0·1100	0
4–12	0	0·2560	0
12–13	0	0·1400	0
12–14	0·1231	0·2559	0
12–15	0·0662	0·1304	0
12–16	0·0945	0·1987	0
14–15	0·2210	0·1997	0
16–17	0·0824	0·1923	0
15–18	0·1070	0·2185	0
18–19	0·0639	0·1292	0
19–20	0·0340	0·0680	0
10–20	0·0936	0·2090	0
10–17	0·0324	0·0845	0
10–21	0·0348	0·0749	0
10–22	0·0727	0·1499	0
21–22	0·0116	0·0236	0
15–23	0·1000	0·2020	0
22–24	0·1150	0·1790	0
23–24	0·1320	0·2700	0
24–25	0·1885	0·3292	0
25–26	0·2544	0·3800	0
25–27	0·1093	0·2087	0
27–28	0	0·3960	0
27–29	0·2198	0·4153	0
27–30	0·3202	0·6027	0
29–30	0·2399	0·4533	0
8–28	0·0636	0·2000	0·0214
6–28	0·0169	0·0599	0·0065

* Impedance and line-charging susceptance in p.u. on a 100000 kVA base
Line charging one-half of total charging of line

Table 13I   Transmission line and transformer data

Bus number	Susceptance* p.u.
10	0·19
24	0·043

* Susceptance in p.u. on a 100000 kVA base

Table 13 J   Static capacitor data

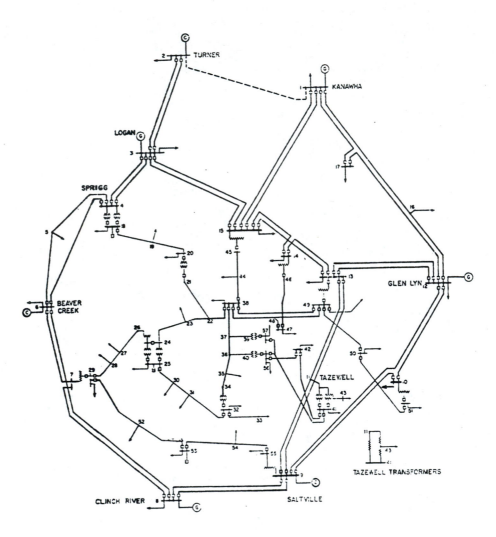

Figure 13C Single-line diagram

IEEE 57-BUS TEST SYSTEM

Line designation	Resistance p.u.*	Reactance p.u.*	Line charging p.u.*
1–2	0·0083	0·0280	0·0645
2–3	0·0298	0·0850	0·0409
3–4	0·0112	0·0366	0·0190
4–5	0·0625	0·1320	0·0129
4–6	0·0430	0·1480	0·0174
6–7	0·0200	0·1020	0·0138
6–8	0·0339	0·1730	0·0235
8–9	0·0099	0·0505	0·0274
9–10	0·0369	0·1679	0·0220
9–11	0·0258	0·0848	0·0109
9–12	0·0648	0·2950	0·0386
9–13	0·0481	0·1580	0·0203
13–14	0·0132	0·0434	0·0055
13–15	0·0269	0·0869	0·0115
1–15	0·0178	0·0910	0·0494
1–16	0·0454	0·2060	0·0273
1–17	0·0238	0·1080	0·0143
3–15	0·0162	0·0530	0·0272
4–18	0	0·555	0
4–18	0	0·43	0
5–6	0·0302	0·0641	0·0062
7–8	0·0139	0·0712	0·0097
10–12	0·0277	0·1262	0·0164
11–13	0·0223	0·0732	0·0094
12–13	0·0178	0·0580	0·0302
12–16	0·0180	0·0813	0·0108
12–17	0·0397	0·1790	0·0238
14–15	0·0171	0·0547	0·0074
18–19	0·4610	0·6850	0
19–20	0·2830	0·4340	0
20–21	0	0·7767	0
21–22	0·0736	0·1170	0
22–23	0·0099	0·0152	0
23–24	0·1660	0·2560	0·0042
24–25	0	1·182	0
24–25	0	1·23	0
24–26	0	0·0473	0
26–27	0·1650	0·2540	0
27–28	0·0618	0·0954	0
28–29	0·0418	0·0587	0
7–29	0	0·0648	0
25–30	0·1350	0·2020	0
30–31	0·3260	0·4970	0
31–32	0·5070	0·7550	0
32–33	0·0392	0·0360	0
32–34	0	0·9530	0
34–35	0·0520	0·0780	0·0016
35–36	0·0430	0·0537	0·0008
36–37	0·0290	0·0366	0
37–38	0·0651	0·1009	0·0010
37–39	0·0239	0·0379	0
36–40	0·0300	0·0466	0
22–38	0·0192	0·0295	0
11–41	0	0·7490	0
41–42	0·2070	0·3520	0
41–43	0	0·4120	0
38–44	0·0289	0·0585	0·0010
15–45	0	0·1042	0
14–46	0	0·0735	0
46–47	0·0230	0·0680	0·0016
47–48	0·0182	0·0233	0
48–49	0·0834	0·1290	0·0024
49–50	0·0801	0·1280	0
50–51	0·1386	0·2200	0
10–51	0	0·0712	0
13–49	0	0·1910	0
29–52	0·1442	0·1870	0
52–53	0·0762	0·0984	0
53–54	0·1878	0·2320	0
54–55	0·1732	0·2265	0
11–43	0	0·1530	0
44–45	0·0624	0·1242	0·0020
40–56	0	1·1950	0
56–41	0·5530	0·5490	0
56–42	0·2125	0·3540	0
39–57	0	1·3550	0
57–56	0·1740	0·2600	0
38–49	0·1150	0·1770	0·0030
38–48	0·0312	0·0482	0
9–55	0	0·1205	0

Table 13K Transmission line and transformer data

Bus number	Starting bus voltage		Generation		Load	
	Magnitude p.u.	Phase angle, deg	MW	MVAr	MW	MVAr
1*	1·04	0	0	0	55·0	17·0
2	1·0	0	0	0	3·0	88·0
3	1·0	0	40	0	41·0	21·0
4	1·0	0	0	0	0	0
5	1·0	0	0	0	13·0	4·0
6	1·0	0	0	0	75·0	2·0
7	1·0	0	0	0	0	0
8	1·0	0	450	0	150·0	22·0
9	1·0	0	0	0	121·0	26·0
10	1·0	0	0	0	5·0	2·0
11	1·0	0	0	0	0	0
12	1·0	0	310	0	377·0	24·0
13	1·0	0	0	0	18·0	2·3
14	1·0	0	0	0	10·5	5·3
15	1·0	0	0	0	22·0	5·0
16	1·0	0	0	0	43·0	3·0
17	1·0	0	0	0	42·0	8·0
18	1·0	0	0	0	27·2	9·8
19	1·0	0	0	0	3·3	0·6
20	1·0	0	0	0	2·3	1·0
21	1·0	0	0	0	0	0
22	1·0	0	0	0	0	0
23	1·0	0	0	0	6·3	2·1
24	1·0	0	0	0	0	0
25	1·0	0	0	0	6·3	3·2
26	1·0	0	0	0	0	0
27	1·0	0	0	0	9·3	0·5
28	1·0	0	0	0	4·6	2·3
29	1·0	0	0	0	17·0	2·6
30	1·0	0	0	0	3·6	1·8
31	1·0	0	0	0	5·8	2·9
32	1·0	0	0	0	1·6	0·8
33	1·0	0	0	0	3·8	1·9
34	1·0	0	0	0	0	0
35	1·0	0	0	0	6·0	3·0
36	1·0	0	0	0	0	0
37	1·0	0	0	0	0	0
38	1·0	0	0	0	14·0	7·0
39	1·0	0	0	0	0	0
40	1·0	0	0	0	0	0
41	1·0	0	0	0	6·3	3·0
42	1·0	0	0	0	7·1	4·4
43	1·0	0	0	0	2·0	1·0
44	1·0	0	0	0	12·0	1·8
45	1·0	0	0	0	0	0
46	1·0	0	0	0	0	0
47	1·0	0	0	0	29·7	11·6
48	1·0	0	0	0	0	0
49	1·0	0	0	0	18·0	8·5
50	1·0	0	0	0	21·0	10·5
51	1·0	0	0	0	18·0	5·3
52	1·0	0	0	0	4·9	2·2
53	1·0	0	0	0	20·0	10·0
54	1·0	0	0	0	4·1	1·4
55	1·0	0	0	0	6·8	3·4
56	1·0	0	0	0	7·6	2·2
57	1·0	0	0	0	6·7	2·0

* Swing machine

Table 13L Bus data

* Impedance and line charging susceptance in p.u. on a 100000kVA base
Line charging: one-half of total charging of line

91

Bus number	Voltage magnitude p.u.	Minimum MVAr capability	Maximum MVAr capability
2	1·01	−17	50
3	0·985	−10	60
6	0·98	−8	25
8	1·005	−140	200
9	0·98	−3	9
12	1·015	−50	155

Table 13M Voltage-controlled (type 2) bus data

Transformer designation	Tap setting*
18–4	0·97
18–4	0·978
29–7	0·967
55–9	0·94
51–10	0·93
41–11	0·955
43–11	0·958
49–13	0·895
46–14	0·9
45–15	0·955
20–21	1·043
25–24	1·000
25–24	1·000
26–24	1·043
32–34	0·975
57–39	0·98
56–40	0·958

Table 13N Tap settings of fixed-tap transformers (The tape is on the 2nd bus number.)

Bus number	Susceptance* p.u.
18	0·1
25	0·059
53	0·063

* Susceptance in p.u. on a 100 000 kVA base

Table 13O Static capacitor data

16. DESIGN PROJECTS

© 1990 IEEE. Reprinted, with permission from:
IEEE Transactions on Power Systems, Vol. 5, No. 4, November 1990

STUDENT DESIGN PROJECTS IN POWER ENGINEERING

J. Duncan Glover
Member, IEEE

Leonard F. Dow
Senior Member, IEEE

ECE Department
Northeastern University, Boston, MA

Boston Edison Company
Boston, MA

Abstract: This paper describes five power engineering projects assigned to juniors, seniors and graduate students at three universities. These projects utilize a personal computer software package that includes transmission line, power flow, short circuit, and transient stability computer programs. The projects, the way they are utilized, and student reaction are discussed.

Keywords: student design projects, power system design, personal computers, software package, power engineering education

INTRODUCTION

A personal computer (PC) software package [1] was introduced in 1987 in junior/senior power engineering courses at Northeastern University-Boston MA. The purposes of the package are to update power curricula; enhance student interest; support classroom instruction; enable students to work on more difficult, practical problems; and initiate student design projects.

The software package includes a set of nine computer programs stored on three diskettes [1]; a software manual [2] including user instructions, sample runs, and flow charts; and an accompanying text [3] which covers the theory and background for the programs. The software, which has been used by students and educators at several universities, is now debugged and user friendly. A complimentary copy of the package is available to instructors who adopt the text for their course. Also, the copyright agreement allows instructors to make copies of the disks for their students.

90 WM 154-5 PWRS   A paper recommended and approved by the IEEE Power Engineering Education Committee of the IEEE Power Engineering Society for presentation at the IEEE/PES 1990 Winter Meeting, Atlanta, Georgia, February 4 - 8, 1990. Manuscript submitted August 30, 1989; made available for printing December 27, 1989.

* The Second Edition of the software requires a PC that can run Microsoft Windows 3.1 or higher.

The programs are written to run on an IBM PC, XT, AT, or any compatible personal computer with DOS 2.1 or higher. The basic requirements are that the computer should have two 360K floppy disk drives, or one floppy disk drive and one hard disk drive, 256K of RAM, and a monochrome monitor. Additionally, if either the power flow, short circuit, or transient stability program is used for a power system with greater than 30 buses, more RAM (expanded to 1 Meg for 100 buses) is required. A printer is also needed for outputs.

The software package was initially introduced at Northeastern University through classroom instruction, including one or two computer sessions with the instructor, and through student PC assignments [1]. Subsequently, instructor time was invested, both at Northeastern and at other universities, to prepare power engineering design projects that employ the software package and have some flexibility in their solutions. The projects have now been tested in classrooms to verify that they can be worked in reasonable amounts of time. Five of these projects, the way they are utilized, and student reaction are discussed here.

POWER FLOW / SHORT CIRCUITS PROJECT

This project is based on a project assigned by Professor Glover during the 1989 Winter quarter in ECE1231, Electric Power Laboratory 1, a one quarter-hour junior/senior laboratory course at Northeastern . A co-requisite for this laboratory is ECE1472, Electrical Power Systems 2, a four quarter-hour course. The project consists of four assignments handed out during the third week of the quarter. The due dates were coordinated with power flow and short circuits material presented in the ECE1472 course. The laboratory grade was based entirely on performance on this project. Each assignment was graded and returned before the next assignment was due.

Time given: 7 weeks
Approximate time required: 30 hours
Weighting of the project towards the lab grade: 100%
Related chapters in text [3]: Chapter 7 - "Power Flow", Chapter 8 - "Symmetrical Short Circuits", and Chapter 9 - " Unsymmetrical Faults "
Additional References: [6,7]
PC Programs used: "Power Flow", "Symmetrical Short Circuits", and "Short Circuits"

Each student is assigned one of the single-line diagrams shown in Figures 1 and 2. Also, the length of line 2 in these figures is varied for each student.

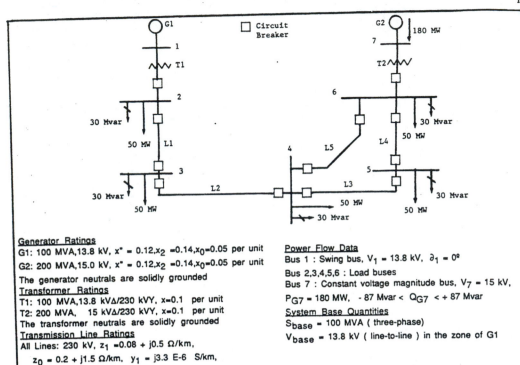

Generator Ratings
G1: 100 MVA, 13.8 kV, x" = 0.12, x_2 =0.14, x_0=0.05 per unit
G2: 200 MVA, 15.0 kV, x" = 0.12, x_2 =0.14, x_0=0.05 per unit
The generator neutrals are solidly grounded
Transformer Ratings
T1: 100 MVA, 13.8 kVΔ/230 kVY, x=0.1 per unit
T2: 200 MVA, 15 kVΔ/230 kVY, x=0.1 per unit
The transformer neutrals are solidly grounded
Transmission Line Ratings
All Lines: 230 kV, z_1 =0.08 + j0.5 Ω/km,
   z_0 = 0.2 + j1.5 Ω/km, y_1 = j3.3 E-6 S/km,
   Maximum MVA = 400
Line Lengths: L_1 = 15 km, L_2 assigned by the instructor (
20 to 50 km), L_3 = 40 km, L_4 = 15 km, L_5 = 50 km.

Power Flow Data
Bus 1 : Swing bus, V_1 = 13.8 kV, ∂_1 = 0°
Bus 2,3,4,5,6 : Load buses
Bus 7 : Constant voltage magnitude bus, V_7 = 15 kV,
P_{G7} = 180 MW, - 87 Mvar < Q_{G7} < + 87 Mvar
System Base Quantities
S_{base} = 100 MVA (three-phase)
V_{base} = 13.8 kV (line-to-line) in the zone of G1

Figure 1   Power Flow/ Short Circuits Project
Single-Line Diagram : Transmission Loop

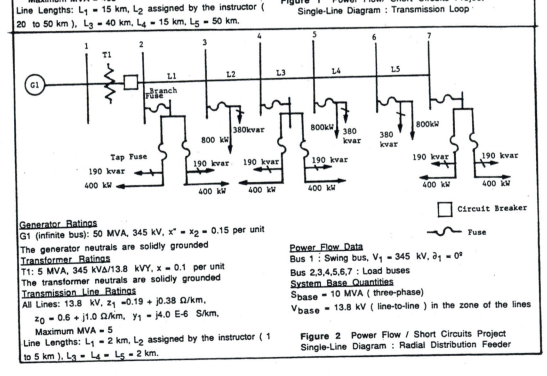

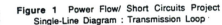

Generator Ratings
G1 (infinite bus): 50 MVA, 345 kV, x" = x_2 = 0.15 per unit
The generator neutrals are solidly grounded
Transformer Ratings
T1: 5 MVA, 345 kVΔ/13.8 kVY, x = 0.1 per unit
The transformer neutrals are solidly grounded
Transmission Line Ratings
All Lines: 13.8 kV, z_1 =0.19 + j0.38 Ω/km,
   z_0 = 0.6 + j1.0 Ω/km, y_1 = j4.0 E-6 S/km,
   Maximum MVA = 5
Line Lengths: L_1 = 2 km, L_2 assigned by the instructor (1
to 5 km), L_3 = L_4 = L_5 = 2 km.

Power Flow Data
Bus 1 : Swing bus, V_1 = 345 kV, ∂_1 = 0°
Bus 2,3,4,5,6,7 : Load buses
System Base Quantities
S_{base} = 10 MVA (three-phase)
V_{base} = 13.8 kV (line-to-line) in the zone of the lines

Figure 2   Power Flow / Short Circuits Project
Single-Line Diagram : Radial Distribution Feeder

Assignment 1: Power Flow Preparation

For the single-line diagram that you have been assigned (Figure 1 or 2), convert all positive-sequence impedance, load and voltage data to per unit using the given system base quantities. Then using the power flow program, create three input data files: bus input data, line input data, and transformer input data. Note that Bus 1 is the swing bus. Your output for this assignment consists of three power flow input data files.

The purpose of this assignment is to get started, and to correct errors before going to the next assignment. It requires a knowledge of the per unit system, which was covered in Electrical Power Systems 1, but may need review.

Assignment 2: Power Flow

Case 1. Run the power flow program and obtain the bus, line and transformer input/output data files that you prepared in Assignment 1.

Case 2. Suggest one method of increasing the voltage magnitude at bus 4 by 5%. Demonstrate the effectiveness of your method by making appropriate changes to the input data of case 1 and by running the power flow program.

Your output for this assignment consists of 12 data files, 3 input and 3 output data files for each case, along with a one-paragraph explanation of your method for increasing the voltage at bus 4 by 5%.

During this assignment, course material contains voltage control methods, including use of generator excitation control, tap changing and regulating transformers, static capacitors, static var systems, and parallel transmission lines.

Assignment 3: Symmetrical Short Circuits

For the single-line diagram that you have been assigned, convert the positive-sequence reactance data to per unit using the given base quantities. For synchronous machines, use subtransient reactance. Then using the symmetrical short circuits program, create the machine, transmission line and transformer input data files. Next run the program to compute subtransient fault currents for a bolted three-phase-to-ground fault at bus 1, then at bus 2, then at bus 3, etc. Also compute bus voltages during the faults and the positive-sequence bus impedance matrix. Assume 1.0 per unit prefault voltage. Neglect prefault load currents and all losses.

Your output for this assignment consists of three input data files and three output data (fault currents, bus voltages, and the bus impedance matrix) files.

Assignment 4: Short Circuits-Breaker/Fuse Selection

For the single-line diagram that you have been assigned, convert the zero-, positive-, and negative-sequence reactance data to per unit using the given system base quantities. Use subtransient machine reactances. Then using the short circuits program, create the generator, transmission line and transformer input data files. Next run the program to compute subtransient fault currents for (1) single - line- to- ground, (2) line-to-line, and (3) double-line-to-ground bolted faults at each bus. Also compute the zero-, positive-, and negative-sequence bus impedance matrices. Assume 1.0 per unit prefault voltage. Also, neglect prefault load currents and all losses.

For students assigned to Figure 1:

Select a suitable circuit breaker from Table 8.10 of the text [3] for each location shown on your single-line diagram. Each breaker that you select should: (1) have a rated voltage larger than the maximum system operating voltage; (2) have a rated continuous current at least 30% larger than normal load current (normal load currents are computed in Assignment 2 above); and (3) have a rated short circuit current larger than the maximum fault current for any type of fault at the bus where the breaker is located (fault currents are computed in Assignments 3 and 4). This conservative practice of selecting a breaker to interrupt the entire fault current, not just the contribution to the fault through the breaker, allows for future increases in fault currents. Note: assume that the (X/R) ratio at each bus is less than 15, such that the breakers are capable of interrupting the dc-offset in addition to the subtransient fault current. Circuit breaker cost should also be a factor in your selection. Do not select a breaker which interrupts 63 kA if a 40 kA or a 31.5 kA breaker will do the job.

For students assigned to Figure 2:

Enclosed [6,7] are "melting time" and "total clearing time" curves for K rated fuses with continuous current ratings from 15 to 200 A . Select suitable branch and tap fuses from these curves for each of the following three locations on your single-line diagram: bus 2, bus 4, and bus 7. Each fuse you select should have a continuous current rating that is at least 15% higher but not more than 50% higher than the normal load current at that bus (normal load currents are computed in Assignment 2). Assume that cables to the load can withstand 50% continuous overload currents. Also, branch fuses should be coordinated with tap fuses; that is, for every fault current, the tap fuse should clear before the branch fuse melts. For each of the three buses, assume a reasonable X/R ratio and determine the asymmetrical fault current for a three-phase bolted fault (subtransient current is computed in Assignment 3). Then for the fuses that you select from [6,7], determine the clearing time CT of tap fuses and the melting time MT of branch fuses. The ratio MT/CT should be less than 0.75 for good coordination.

TRANSMISSION LINE DESIGN PROJECT

This is based on a project assigned by Professor Herman Hill Jr. during the Fall 1987 quarter in EE455, Introduction to Electric Power System Engineering and Analysis, a 4 quarter-hour senior course at Ohio University, Athens, OH. An augmented version was also assigned in EE555, a graduate power system engineering course.

Time given: 6 weeks
Approximate time required: 25 hours
Weighting of the project towards the course grade: 20%
Related chapter in text [3]: Chapter 5 - "Transmission
  Line Parameters"
Additional References: [4, 5]
PC Program used: "Line Constants"

Project Description - EE455

Design a three-phase, 60 Hz transmission line rated to continuously transfer 1000 MVA. Specifically, you are to provide:

(1) A detailed sketch showing placement of phase conductors and shield wire(s), their height above ground both at towers and at the point of maximum sag between towers, the average span, and the right-of-way required (see Figure 5.31 of the text [3] for an example of the information required).

(2) A description of the composition and construction of phase conductors and shield wire(s) (see Table A.4 of the text [3]).

(3) The following transmission line electrical parameters:
(3.1) Rated voltage (kV)
(3.2) Rated current (A)
(3.3) Series sequence impedance matrix (Ω/km)
(3.4) Shunt sequence admittance matrix (S/km)
(3.5) Conductor surface electric field
      strength (kV/cm)
(3.6) Lateral profile of ground-level electric
      field strength (kV/m)

(4) A one-page typed description (with decent prose, good grammar, and correct spelling) of the advantages and disadvantages of your design. Your discussion should include technical factors (such as line losses, electric field strengths, clearances, etc.) and economic factors (including capital cost of installation and operating cost).

Assume a 75 km line length, level terrain, and 100 Ω-m earth resistivity. Current ratings are to be based on conductor thermal limits only; whereas, ratings of terminal equipment, such as circuit breakers, and system issues such as stability and voltage regulation are not to be considered here. Also, the composition of the towers and structural issues are not part of this project.

Grades will be based on the completeness of the project, practicality of the design, and quality of the one-page description. There is no good reason why your design should closely resemble either the text examples or the design of any other student.

Project Description - EE555

Graduate students shall do the same project as described above, except that:

(1) Your line must be a double-circuit line
(2) Your phase conductors must be bundled
(3) Your design and discussion must include corona considerations.

Additional Assignment

An additional assignment is given (not done in EE455/55) as follows:

Additional time given: 6 weeks
Additional time required: 25 hours
Weighting of the project towards the course grade: 20%
Related chapter in text [3]: Chapter 6- "Transmission
  Lines: Steady-State Operation"
PC Program used: "Transmission Lines: Steady State
  Operation"

Evaluate your line design for the following three line lengths: 75, 200, and 500 km. Assume steady-state, positive-sequence operation. For each length, determine:

(1) ABCD parameters (with no line compensation)
(2) Equivalent pi circuit series impedance (Ω) and shunt admittance (S) (with no compensation)
(3) The total amount of shunt reactive compensation (%) required to achieve 10 % voltage regulation for this line (full load is 1000 MVA at unity power factor). One-third of the total compensation is to be placed at the sending end, at the center, and at the receiving end of the line. Shunt compensation is removed at heavy loadings.
(4) ABCD parameters and equivalent pi circuit for the shunt compensated line
(5) Maximum theoretical power (MW) that the line can transfer at rated sending- and receiving-end voltages.
(6) Practical line loadability (MW) based on a voltage drop limit $V_R/V_S \geq$ 0.95 and 35º maximum angular displacement across the line.

POWER FLOW/TRANSIENT STABILITY PROJECT

This is based on a project assigned by Adjunct Professor Richard Farmer during the Spring 89 semester in EEE471, Power System Analysis, a 3 semester-hour senior course at Arizona State University, Tempe, AZ. This course can also be taken by graduate students for graduate credit. The project consists of two assignments, a power flow assignment and a transient stability assignment.

Time given: 6 weeks
Approximate time required: 50 hours
Weighting of the project towards the course grade: 40%
Related chapters in text [2]: Chapter 7 - "Power Flow",
    and Chapter 12 - "Transient Stability"
PC Programs used: "Power Flow" and "Transient Stability"

Project Description

Figure 3 shows the single-line diagram of a 10-bus power system with 7 generating units, 2 345-kV lines, 7 230-kV lines, and 5 transformers. Per unit transformer leakage reactances, transmission line series impedances and shunt susceptances, real power generation, and real and reactive loads during heavy load periods, all on a 100 MVA system base, are given on the diagram. Fixed transformer tap settings are also shown. During light load periods, the real and reactive loads (and generation) are 25% of those shown. Note that bus 1 is the swing bus.

Power Flow Assignment

Using the power flow program (convergence can be achieved by changing load buses to constant voltage magnitude buses with wide var limits) determine:

(1) The amount of shunt compensation required at 230 and 345 kV buses such that the voltage magnitude $0.99 \leq V \leq 1.02$ per unit at all buses during both light and heavy loads. Find two settings for the compensation, one for light and heavy loads.

(2) The amount of series compensation required during heavy loads on each 345-kV line such that there is a maximum of 40° angular displacement between bus 4 and bus 10. Assume that one 345-kV line is out of service. Also assume that the series compensation is effectively distributed such that the net series reactance of each 345-kV line is reduced by the % compensation. Determine the % series compensation to within ± 10%.

Transient Stability Assignment

Transient reactances X_d' and H constants, both in per unit based on generator ratings, are given as follows:

Unit	MVA Rating	X_d'	H
1	333	0.30	3.8
2	333	0.30	3.8
3	210	0.25	3.5
4	210	0.25	3.5
5	210	0.25	3.5
6	353	0.35	4.2
7	353	0.35	4.2

Convert the above data to per unit on a 100 MVA system base and create the synchronous machine input data file in the Transient Stability program. The above 7 units may be separately entered into the input data file, or those units which are connected to the same bus may be combined to enter 4 equivalent units into the data file.

Use the Transient Stability program to find the percent series compensation required on each 345-kV line to maintain stability for the following event:

t=0 : A bolted three-phase fault at bus 4 (on one 345-kV line from bus 4 to bus 10)
t=5 cycles (0.083333 sec): The fault is cleared by permanently removing the faulted 345-kV line from bus 4 to bus 10.

Find the % series compensation to the nearest 10%. Assume a heavy load period with both 345 kV lines initially in service.

PLANNING PROJECT

This is based on a project assigned by Adjunct Professor Leonard Dow during the Spring 89 quarter in ECE3412, Power System Planning, a 4 quarter-hour graduate course at Northeastern. This course can also be taken by seniors with special permission.

Time given: 11 weeks
Approximate time required: 40 hours
Weighting of the project towards the course grade: 25%
Related chapters in text [3]: Chapter 7 - "Power Flow"
Additional References: [8,9]
PC Program used: "Power Flow"

Project Description

Figure 4 shows a single-line diagram of four interconnected power systems identified by different graphic bus designations . The following data are given:

(1) There are 31 buses, 21 lines, and 13 transformers.
(2) Generation is present at buses 1, 16,17,22, and 23.
(3) Total load of the four systems is 400 MW.
(4) Bus 1 is the swing bus.
(5) The system base is 100 MVA
(6) Additional information on transformers and transmission lines is provided in [8,9].

Based on the data given:

(1) Allocate the total 400 MW system load among the four systems.
(2) For each system, allocate the load to buses that you want to represent as load buses. Select reasonable load power factors.
(3) Taking into consideration the load you allocated above, select appropriate transmission line voltage ratings, MVA ratings, and distances necessary to supply these loads. Then determine per unit transmission line impedances for the lines shown on the single-line diagram (show your calculations).

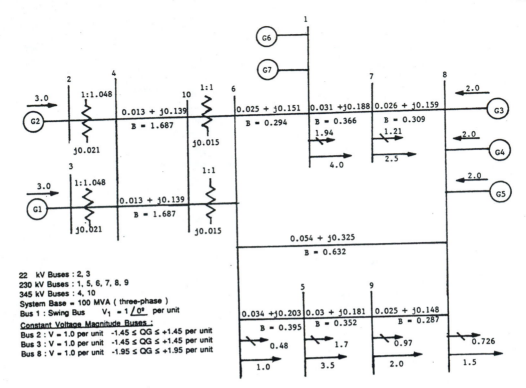

Figure 3 Power Flow/Transient Stability Project Single-Line Diagram: 10 Bus Power System

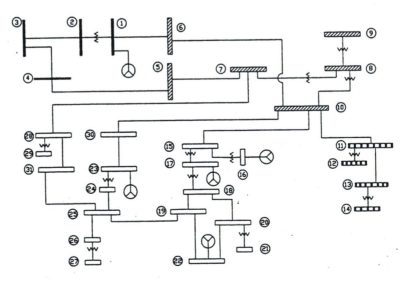

Figure 4 Planning Project Single-Line Diagram: 31 Bus Interconnected Power System

(4) Also select appropriate transformer voltage and MVA ratings, and determine per unit transformer leakage impedances for the transformers shown on the single-line diagram.

(5) Develop a generation schedule for the 5 generator buses.

(6) Show on a copy of the the single-line diagram per unit: line impedances, transformer impedances, generator outputs and loads that you have selected above.

(7) Using the Power Flow program, run a base case power flow. In addition to the printed input/output data files, show on a separate copy of the single-line diagram per unit bus voltages as well as real and reactive line flows, generator outputs, and loads. Flag any high/low bus voltages for which $0.95 \leq V \leq 1.05$ per unit and any line or transformer flows that exceed normal ratings.

(8) If the base case shows any high/low voltages or ratings exceeded, then correct the base case by making changes. Explain the changes you have made.

(9) Repeat (7) above. Rerun the Power Flow program and show your changes on a separate copy of the single-line diagram.

(10) Provide a typed summary of your results along with your above calculations, printed power flow input/output data files, and copies of the single-line diagram.

TRANSIENTS / TRANSIENT STABILITY PROJECT

This project is based on a project assigned by Professor Glover during the 1989 Spring quarter in ECE1379, Transients in Electric Power Systems, a four quarter-hour junior/senior course at Northeastern. The project consists of five assignments handed out during the third week of the quarter.

Time given: 7 weeks
Approximate time required: 30 hours
Weighting of the project towards the course grade: 20%
Related chapters in text [3]: Chapter 11 - "Transmission Lines: Transient Operation", and Chapter 12 - "Transient Stability"
PC programs used: "Transmission-Line Transients", "Power Flow," and "Transient Stability"

Assignment 1: Transients

Figure 11.3 of the text [3] shows a single-phase two-wire lossless line with source and load terminations. The voltage source at the sending end is a 100 kV step with internal impedance Z_G. The line has a 30 km length, characteristic impedance $Z_C = 250 \ \Omega$, and a 100 μs transit time. Use the Transmission-Line Transients program to compute the voltages for $0 \leq t \leq 1000$ μs at the sending end, center, and receiving end of the line for examples 11.2, 11.3, and 11.4 of the text [3], given as follows:

Example 11.2 $Z_G = Z_C$, and the receiving end is open.

Example 11.3 $Z_G = Z_C$, and the receiving end is terminated by a 1 μF capacitor.
Example 11.4 $Z_G = 2Z_C$, and $Z_R = Z_C/3$ at the receiving end.

Use the following calculation data: Calculation Time = 1 μs, Final Time = 1000 μs, and Printout Time = 20 μs. Also, select 10 S conductance for the series element at each end of your line sections, which approximates a lossless line (∞ cannot be entered).

Assignment 2: Transients

Figure 11.10 of the text [3] shows a single-phase lossless line A with characteristic impedance Z_A, length L_A, and transit time t_A terminated by single-phase lossless line B with characteristic impedance Z_B, length L_B, and transit time t_B. Use the Transmission-Line Transients program to compute the sending end, junction, and receiving end voltages for $0 \leq t \leq 1000$ μs for the following three examples.

(1) The source at the sending end of line A is a 100 kV step with internal impedance $Z_G = Z_A$. The load at the receiving end of line B is $Z_R = 2Z_B$. The line data are $Z_A = 400 \ \Omega$, $L_A = 30$ km, $Z_B = 100 \ \Omega$, $L_B = 20$ km, and, $t_A = t_B = 100$ μs (see Example 11.6 of the text [3]). Select 10 S conductance for the series elements at the ends of line A and B, approximating lossless lines.

(2) Repeat (1) above except that the lines are lossy with series resistance 0.5 Ω/km. Thus, select 0.1333 S and 0.2 S conductance at each end of line A and B, respectively, to represent line losses. Give a one paragraph comparison of the results of (1) and (2).

(3) The source at the sending end of line A is a 100 kV pulse of duration 100 μs with internal impedance $Z_G = Z_A$. The receiving end of line B is short circuited. The line data are $Z_A = Z_B = 400 \ \Omega$, $L_A = L_B = 50$ km, and, $t_A = t_B = 166.7$ μs. Also, a 400 Ω shunt resistor is installed at the junction of line A and B (see Example 11.6 of the text [3]). Select 10 S conductance for the series elements at the ends of line A and B, approximating lossless lines. Also, select 100 S conductance for the shunt branch at the end of line B, approximating a short circuit.

Use the same calculation data as in Assignment 1.

Assignment 3: Transients

As described in Example 11.9 of the text [3], lightning strikes the center of a single-phase, unshielded distribution line that is terminated at one end by a transformer and at the other end by an open circuit. The line is initially unenergized. Run the Transmission-Line Transients program to compute the voltage at the center and at both ends of the line for $0 \leq t \leq 150$ μs for the following two cases.

Case 1. Use the following data:
Line: 20 kV, 10 km, 300 Ω characteristic impedance, 33.33 μs transit time, 0.05 Ω/km series line losses.
Transformer: 10 MVA, 20/8 kV, 200 kV BIL, 0.006 μF winding capacitance-to-ground.
Lightning surge: An ideal, 20 kA square-wave current source with 20 μs duration.
Calculation Time Interval = 0.1 μs, Print Interval = 3 μs

Case 2. Repeat (1) with a 21 kV surge arrester installed adjacent to the transformer. The arrester is modeled by a piecewise constant resistor with 0.5 μS conductance when the arrester voltage is below 55 kV, and 0.222 S when the arrester voltage exceeds 55 kV. Verify that the arrester holds the transformer voltage to 88.6 kV, well below the transformer BIL.

Assignment 4: Transient Stability

Figure 10(a) of the software manual [2] shows a 5 bus power system with 6 lines, 2 transformers, and 2 generators. Run the Transient Stability program for a three-phase bolted fault on Line 1 at Bus 2. For prefault conditions, use the power flow data from the sample run given in the manual. Synchronous machine data is also given in the manual. Run the following two cases:

Case 1. The fault is cleared after 0.5 seconds by permanently removing Line 1.

Case 2. Same as (1), except that the fault is cleared after 0.54 seconds.

Use a 0.005 second integration interval, 0.02 second printout interval, and 1.0 second final time. For each run, plot the angle of generator 1 and 2 versus time, using the output data. Show that Run (1) is stable and Run (2) is unstable. Therefore, the critical clearing time for this fault is $0.50 \leq t_{crit} \leq 0.54$ seconds.

Assignment 5: Transient Stability

Lines 5 and 6 are initially out of service for the power system given in Assignment 4. A three-phase bolted fault occurs on Line 1 at Bus 2, which is then cleared by permanently removing Line 1. Determine the critical clearing time in seconds and in cycles (at 60 Hz).

First rerun the Power Flow program with Lines 5 and 6 removed, to obtain the new prefault conditions. Then run the Transient Stability program with the same data as in Assignment 4, except that you will have to select the fault clearing time. After making a series of runs, obtain one run that is stable and one run that is unstable, with no more than 0.01 seconds between the two fault clearing times. In this way, you can determine t_{crit} within 0.01 seconds (0.6 cycles).

REACTION

Student reaction to the projects has been generally positive. Using the software to solve practical problems has tied the course material more closely to problems encountered by industry professionals, and has also enhanced student understanding of the various power system topics presented in the classroom.

Students who completed the power flow-short circuits project found the circuit breaker-fuse selection particularly interesting; they understood the value of using the software to obtain normal and short-circuit currents.

Some of the undergraduate students who completed the transmission line design project had difficulty. Since there was a range of acceptable solutions, with no formula to plug in to get "the answer," they were uneasy about their designs. Their inexperience, both with transmission lines and with design oriented projects in general, caused their uneasiness. Hopefully, their completion of this design will give them more confidence with the next one.

The planning project enabled students to study the impact of load growth on an existing system, and to simulate different generation and/or transmission plans to provide for the increased load. The project was aided by part-time graduate students from industry who were familiar with power flow preparation. Sharing of their experiences in the classroom resulted in more realistic power flow solutions by other students.

Some students had difficulty getting started on the power flow/transient stability project. They were unfamiliar with the power flow program and its input data. After learning how to use the program, however, they were able to complete the project. Typically, each student ran the power flow program about 30 times in order to obtain 10 successful runs to be submitted.

Our reaction has also been positive. The design projects effectively augment classroom lectures and homework, to the extent that students have a stronger understanding of the theory. We found that the software is now debugged and sufficiently user friendly, such that minimal student effort and instructor support are required to execute programs. After reviewing student work, we have found a need to emphasize quality in technical report writing. In addition to computer printouts and calculation sheets, students should be required to submit typed project summaries with decent prose, good grammar, and correct spelling, in order to adequately describe their designs.

CONCLUSION

The availability of the personal computer software package has extended classroom capabilities such that student design projects similar to the five · described here can be incorporated into power engineering courses and/or laboratories. The package has enabled students to work on more difficult problems and has become an innovative tool in the learning process. Other educators are welcome to use these projects for their courses, and are encouraged to use the ideas presented here to develop their own projects. The nine programs that are stored on diskettes can also be utilized by students later in their careers and by professionals.

REFERENCES

[1] J. D. Glover, " A Personal Computer Package for Power Engineering Education", *IEEE Transactions on Power Systems* , Volume 3, No. 4 , pp. 1864–72, November, 1988.

[2] J. D. Glover, *Power System Analysis and Design Software* , Boston: PWS Publishers, 1987.

[3] J. D. Glover and M.S. Sarma, *Power System Analysis and Design with Personal Computer Applications* , Boston: PWS Publishers, 1987.

[4] ANCI C2.1987, *National Electrical Safety Code* , New York: Institute of Electrical and Electronics Engineers, 1984 edition.

[5] Electric Power Research Institute (EPRI), *Transmission Line Reference Book : 345 kV and Above* , Palo Alto, CA: EPRI, 2nd edition, 1982.

[6] McGraw Edison Company, *Fuse Catalog* , R240-91-1, Canonsburg, PA: Mcgraw Edison, April, 1985.

[7] Westinghouse Electric Corporation, *Electric Utility Engineering Reference Book : Distribution Systems,* Pittsburgh: Westinghouse, 1959.

[8] Westinghouse Electric Corporation, *Transmission and Distribution Reference Book* , Pittsburgh: Westinghouse, 4th edition, 1964.

[9] Aluminum Association, *The Aluminum Electrical Conductor Handbook* , Washington DC : Aluminum Association, 1971.

ACKNOWLEDGMENT

The authors are grateful to Herman Hill Jr.- Professor at Ohio University, Athens, OH; and to Richard Farmer-Adjunct Professor at Arizona State University, Tempe, AZ ; for sharing their projects, described herein. Making these projects available to the educational community certainly enhances power engineering education.

BIOGRAPHIES

J. Duncan Glover was born in Melrose, MA on November 4, 1944. He received the BSEE degree Magna Cum Laude from U. Mass-Amherst in 1966, and the MSEE and Ph.D. degrees from MIT-Cambridge, MA in 1968 and 1971, respectively. He worked as an engineer in EHV planning at the American Electric Power Service Corporation, NY, NY from 1971 to 1973, and as a senior engineer in HVDC transmission design at Commonwealth Associates, Jackson, MI in 1974. From 1974-1976, he was a consulting engineer with the International Engineering Company on the ITAIPU hydroelectric power project, Rio de Janeiro, Brazil. Since 1977, he has been at Northeastern University, where he is currently Associate Professor of Electrical and Computer Engineering. His research interests include EHV design, power system dynamics, and automatic controls. He is a member of the IEEE and is a Massachusetts registered professional engineer.

Leonard F. Dow is employed by Boston Edison Company, where he has held various engineering and staff positions during his 25 year career. He has been Adjunct Professor at Northeastern University's Graduate School of Engineering in Boston, MA, since 1971. A senior member of the IEEE, Mr. Dow is currently the national Treasurer of the Power Engineering Society. He received his BSEE and MS degrees from Northeastern University in 1963 and 1967, respectively. Mr. Dow was recently honored by both the professional and academic communities. He received the Lawrence Cleveland Award for dedicated service to the engineering profession and the Boston Chapter of the Power Engineering Society, of which he is a past chairman. At Northeastern University, Mr. Dow was presented with the Teacher Excellence Award for the 1988-1989 academic year.

Discussion

B. Jeyasurya (Indian Institute of Technology, Bombay, India): The authors must be commented for presenting some interesting applications of the personal computer software package for power engineering design projects. This approach will certainly enhance the students' understanding of power system topics and relate them to the practical problems. The comments here are specifically related to the Transient Stability Projects. In order to determine the series compensation required to maintain stability or the critical clearing angle with respect to various fault conditions, the students have to run the transient stability programs a large number of times by changing the input data (reactive compensation, clearing time etc.). The stability of the power system can be determined only after analysing the large amount of data as provided by the output file for each run.

In the discusser's opinion, incorporating a suitable Graphics software to plot the variation of machine power angle with respect to time will enhance the usefulness of the transient stability program. By analysing the plot of the power angle one can quickly determine whether the system will remain stable or not for a particular operating condition with respect to a specific contingency. The influence of the location of the fault, clearing time etc. on the dynamic performance of the power system as well as the coherency of some generators [A] can be easily understood by analysing the swing curve. This will enable the users of this package to gain better insight into the transient stability of a power system.

Reference

[A] 'Rapid Analysis of Transient Stability,' IEEE Publication No. 87TH0169-3-PWR, 1987.

Manuscript received March 6, 1990.

J. DUNCAN GLOVER and LEONARD F. DOW: We agree with Dr. Jeyasurya that using graphics software to plot transient stability outputs will enhance the usefulness of the transient stability program. Stability can be evaluated more rapidly from output plots rather than from output data tables, especially when many cases with varying input data need to be run.

User friendliness was of major importance in the development of the PC software package used for design projects described in this paper. We wanted students and other users to learn to run a power flow, short circuit, or transient stability program in a few minutes. As such, we did not incorporate into the package all features that are included in production programs used in the electric utility industry.

A second edition of the software package and accompanying text is currently scheduled for publication in 1992. We plan to include in this new edition more graphics capabilities such as that described by Dr. Jeyasurya.

Manuscript received March 29, 1990,